# Baofeng Radio

A Green Beret's Guide to Master Your Communication Skills,
Ensure Safety and Elevate Emergency

**Tristan Grimes**

Warning – Disclaimer

The purpose of this book is to educate and entertain. The author or publisher does not guarantee that anyone following the techniques, suggestions, tips, ideas or strategies will have successful. The author and publisher shall have neither liability or responsibility to anyone with respect to any loss or damage caused or alleged to be caused, directly or indirectly, by the information contained in this book.

# QR CODE

Unlock additional insights and expertise! Scan the QR code below to download three exclusive bonus guides:

- Decode & Conquer – A Green Beret's Companion To Q-Codes
- Wild Radio Connections – A Green Beret's Companion to Outdoor Communication
- Wilderness Mastery – A Green Beret's Guide to Bushcraft and Advanced Survival Skills

Elevate your Baofeng radio experience and expand your knowledge of outdoor communication and survival skills. Don't miss out on this opportunity to enhance your journey into the world of radios and wilderness mastery.

# Table of Contents

Introduction ................................................................................................................ 7

WELCOME TO THE WORLD OF BAOFENG RADIOS ......................................................... 7

*Chapter 1:* **Getting Started with Baofeng Radios** .......................................... 10

UNBOXING AND FAMILIARIZATION ............................................................................ 10

BASIC COMPONENTS AND FEATURES ........................................................................ 11

SAFETY PRECAUTIONS AND BEST PRACTICES............................................................ 13

POPULAR BAOFENG MODELS .................................................................................... 15

    *Baofeng BF-F8HP* ............................................................................................ 16

    *Baofeng BF-888S* ............................................................................................ 17

    *Baofeng UV-82* ............................................................................................... 17

    *Baofeng DM-5R* .............................................................................................. 18

    *Baofeng GT-3* .................................................................................................. 19

    *Baofeng UV-5R - The Iconic Choice* ............................................................... 20

*Chapter 2:* **Understanding Radio Fundamentals** .......................................... 25

WHAT IS RADIO COMMUNICATION? .......................................................................... 25

FUNDAMENTAL CONCEPTS ...................................................................................... 29

RADIO COMMUNICATION TERMINOLOGY .................................................................. 32

*Chapter 3:* **Setting Up Your Baofeng Radio** .................................................. 36

INITIAL BATTERY CHARGING ................................................................................... 36

ATTACHING THE ANTENNA ...................................................................................... 38

POWERING ON AND OFF .......................................................................................... 39

*Chapter 4:* **Navigating the Baofeng Radio Menu** .......................................... 42

DISPLAY AND BUTTONS OVERVIEW .......................................................................... 42

PROGRAMMING YOUR FIRST CHANNEL ..................................................................... 44

ADJUSTING VOLUME AND SQUELCH ......................................................................... 46

*Chapter 5:* **Basic Radio Operations** .............................................................. 48

TRANSMITTING AND RECEIVING............................................................................... 48

ADJUSTING TRANSMIT POWER................................................................................. 52

DUAL WATCH AND DUAL RECEIVE .......................................................................... 57

USING CTCSS AND DCS TONES ............................................................................... 61

*Chapter 6:* **Programming Your Baofeng Radio** ............................................. 67

PROGRAMMING SOFTWARE AND CABLES................................................................... 67

    *Programming Software* ..................................................................................... 67

    *Programming Cables* ........................................................................................ 73

BASIC PROGRAMMING STEPS ................................................................................... 76

STORING AND MANAGING CHANNELS........................................................................ 78

SETTING UP SCAN LISTS .......................................................................................... 81

USING CHIRP SOFTWARE ....................................................................................................... 84

*Chapter 7:* **Making Your First Contact** ..................................................................................**89**

FINDING LOCAL REPEATERS ................................................................................................ 89

CALLING AND RESPONDING TO CQ ...................................................................................... 94

PROPER RADIO ETIQUETTE.................................................................................................. 98

*Chapter 8:* **Troubleshooting and Maintenance** .................................................................**103**

COMMON ISSUES AND SOLUTIONS....................................................................................... 103

BATTERY CARE AND REPLACEMENT...................................................................................... 108

FIRMWARE UPDATES AND RESETS......................................................................................... 110

*Firmware Updates*........................................................................................................... *110*

*Resets* ............................................................................................................................. *112*

*Chapter 9:* **Safety and Emergency Communications** ........................................................ **114**

USING BAOFENG RADIOS IN EMERGENCIES ......................................................................... 114

CHOOSING THE RIGHT FREQUENCY ..................................................................................... 116

ACCESSING NOAA WEATHER CHANNELS ............................................................................ 119

EMERGENCY CODES AND SIGNALS ....................................................................................... 121

COMMUNICATION PROTOCOLS ............................................................................................ 123

TIPS FOR SIGNAL RESILIENCE ............................................................................................. 125

SEARCH AND RESCUE OPERATIONS ...................................................................................... 128

*Chapter 10:* **Staying Safe and Legal** .....................................................................................**132**

LICENSING REQUIREMENTS ................................................................................................. 132

OPERATING WITHIN THE LAW ............................................................................................. 135

PENALTIES AND ENFORCEMENT ........................................................................................... 137

**Conclusion** .................................................................................................................................**140**

# Introduction

## Welcome to the World of Baofeng Radios

In an interconnected world, communication stands as the cornerstone of progress and collaboration. From casual conversations to critical transmissions during emergencies, the need for reliable, accessible, and versatile communication tools is undeniable. It is within this realm that Baofeng radios emerge as a beacon of innovation, redefining the landscape of handheld communication devices.

Welcome to the intricate world of Baofeng radios, where functionality meets affordability, and versatility merges with reliability. Baofeng, a renowned name in the realm of radio communication, has garnered attention for its range of handheld radios that cater to diverse needs and scenarios. Whether you're an outdoor enthusiast seeking a reliable means to stay connected amidst nature's expanse, an emergency responder requiring a dependable tool in crises, or a professional navigating the demands of different industries, Baofeng radios have something to offer.

## Why Baofeng Radios?

Baofeng radios have carved their niche by offering a unique blend of functionality, affordability, and adaptability. They're not just communication devices; they represent a gateway to seamless connectivity in various scenarios. Their allure lies in their ability to cater to diverse needs, be it for outdoor enthusiasts exploring remote terrains, emergency responders navigating crisis situations, or professionals seeking efficient communication solutions.

What sets Baofeng radios apart is their versatility. They are not confined by the limitations of traditional communication devices. Instead, they stand as a testament to innovation, boasting features that transcend conventional boundaries. Baofeng radios are not just tools; they are enablers, empowering individuals and organizations to bridge distances and communicate effectively, regardless of the circumstances.

## What to Expect from This Guide

This comprehensive guide is a roadmap into the intricate world of Baofeng radios. It is meticulously crafted to be your companion in understanding, utilizing, and maximizing the potential of these remarkable devices.

Within these pages, expect to embark on a journey that starts with the basics and leads you through the complex yet fascinating realm of radio communication. You'll unravel the technical aspects, decode the programming intricacies, and gain insights into the advanced functionalities that make Baofeng radios a formidable choice.

This guide is designed for everyone, whether you're a novice exploring the realm of radios for the first time or an experienced user seeking to delve deeper into the capabilities of Baofeng radios. It presents information in a structured, accessible manner, guiding you through setup procedures, programming techniques, troubleshooting methods, and real-world applications.

Beyond the technicalities, this guide aims to capture the essence of why Baofeng radios have garnered a loyal following. It explores their adaptability across various scenarios, their role in enhancing communication, and their impact on transforming how we connect in the modern world.

Additionally, as a special bonus, you'll receive three exclusive guides:

1. **Decode & Conquer. A Green Beret's Companion To Q-Codes**:

Unravel the mysteries of radio communication with this guide to Q-Codes. Designed for both beginners and seasoned operators, this companion provides essential insights into decoding and using Q-Codes effectively.

2. **Wild Radio Connections. A Green Beret's Companion to Outdoor Communication**:

Navigate the complexities of outdoor communication with this comprehensive guide. From choosing the right frequencies to optimizing antenna placement, this companion is your key to establishing robust radio connections in the wilderness.

3. **Wilderness Mastery. A Green Beret's Guide to Bushcraft and Advanced Survival Skills**:

Elevate your survival skills with this guide from a former Green Beret. Explore advanced bushcraft techniques, sharpen your outdoor navigation skills, and gain insights into handling medical emergencies in the wilderness.

Download these exclusive bonus guides to complement your Baofeng Radio handbook and take your outdoor communication and survival knowledge to the next level!

**Scan the QR code at the beginning of this book!**

Prepare to unlock the potential of Baofeng radios as you journey through this guide. By the end, you'll not only grasp the technical intricacies but also comprehend the broader implications and applications these devices offer.

Welcome aboard this expedition into the world of Baofeng radios, where communication knows no bounds, and possibilities are limitless. Get ready to experience a paradigm shift in the way you perceive and utilize communication tools.

# *Chapter 1:* **Getting Started with Baofeng Radios**

## Unboxing and Familiarization

When unboxing a Baofeng radio, it's essential to carefully inspect each component and ensure that everything is present and undamaged. The satisfaction of unboxing starts with observing the radio, antenna, battery pack, charger, belt clip, and the instruction manual neatly organized within the packaging.

Begin by examining the radio body, which serves as the core of the communication device. Its sturdy build, ergonomic design, and strategically placed controls indicate durability and ease of use. The antenna, a vital component for signal transmission and reception, should be securely attached to the designated port on the radio body.

The battery pack, essential for powering the device, requires careful attention. If not pre-installed, follow the provided instructions to insert the battery correctly into the radio body. Subsequently, connect the charger to the designated port on the radio to commence the initial charging process.

The belt clip, often an accessory included with Baofeng radios, offers convenient portability. Attach it securely to the radio body, ensuring ease of carrying while always keeping the device within reach.

The instruction manual, a crucial component, serves as your comprehensive guide to understanding the functionalities, operation, programming, and troubleshooting of the radio. Keep it accessible for reference throughout the setup process and beyond.

### Acquainting Yourself with the Device

After unpacking the components, it's time to become acquainted with the Baofeng radio:

1. **Physical Examination:** Thoroughly inspect each component for any visible damage incurred during shipping. It's essential to ensure that the components are intact and free from defects.

2. **Battery Installation:** Follow the instructions provided in the manual to correctly install the battery into the radio body. Careful installation ensures optimal functionality and avoids potential issues during usage.

3. **Attaching the Antenna:** Securely attach the antenna to the designated port on the radio body. A properly connected antenna is pivotal for optimizing signal transmission and reception, thereby enhancing communication range and clarity.

4. **Exploring Physical Controls:** Familiarize yourself with the layout and functionality of the physical controls on the radio body. Identify the power button, volume knob, channel selector, PTT (Push-To-Talk) button, and any auxiliary buttons or ports present.

5. **Powering On:** Activate the radio by pressing and holding the power button. As the device powers up, observe the startup sequence displayed on the screen. This sequence often includes model information and default settings.

## Basic Components and Features

Baofeng radios offer a range of features and functionalities. Understanding the basic components and features of these radios is essential for effective use. Let's explore them in detail:

### Core Components

1. **Radio Body:** The radio body serves as the central unit housing the essential components and controls necessary for communication. It encompasses the circuitry, display screen, keypad, speaker, microphone, and antenna connector, creating a cohesive interface for users to operate the device.

2. **Antenna:** An integral part of the radio, the antenna is responsible for transmitting and receiving radio signals. Its design and placement significantly impact the radio's communication range and signal clarity. Users can upgrade or replace antennas for specific purposes or enhanced performance.

3. **Battery Pack:** The battery pack powers the radio, providing the necessary energy for uninterrupted operation. Baofeng radios often use rechargeable lithium-ion batteries, offering users the convenience of recharging for extended use.

4. **Charger:** The charger is essential for replenishing the battery's power. It connects to the radio body, ensuring that the battery remains charged and ready for use whenever needed. Baofeng radios typically come with dedicated chargers to fit the specific battery model.

5. **Belt Clip:** A practical accessory, the belt clip enables users to carry the radio conveniently. It attaches securely to the radio body, allowing users to clip the device onto belts, straps, or pockets for easy access and portability.

**Key Features and Controls**

1. **Power Button:** The power button activates and deactivates the device. Users press and hold this button to turn the radio on or off, initiating or shutting down its operations.

2. **Channel Selector:** Baofeng radios offer multiple channels, frequencies, and memory slots for users to select from. The channel selector enables users to navigate through available frequencies or pre-programmed channels for communication.

3. **Volume Knob:** The volume knob regulates the audio output level, allowing users to adjust the volume according to their preferences and surrounding conditions.

4. **PTT Button (Push-To-Talk):** A fundamental control, the PTT button enables users to initiate transmission when pressed while speaking into the microphone. Releasing the button switches the device back to receive mode.

5. **Menu System:** Baofeng radios feature a menu system accessible through the keypad, providing access to various settings, functions, and customization options. Users can navigate through the menu to configure preferences, set frequencies, enable features, and personalize the device according to their needs.

6. **LCD Display Screen:** The display screen presents essential information, such as channel frequency, signal strength, battery status, and menu navigation. It provides users with real-time feedback about the radio's operations and settings.

7. **Microphone and Speaker:** The microphone allows users to transmit their voice, while the speaker produces audio output for received transmissions. Clear and efficient communication relies on the quality and functionality of these components.

Each component and feature within Baofeng radios serves a specific purpose, collectively contributing to seamless communication. The integration of these elements allows users to operate the device, establish connections, and transmit and receive messages effectively.

Users can explore additional functionalities, such as dual-band operation, scanning modes, squelch settings, CTCSS/DCS tones, and programmable options. These advanced features offer flexibility and customization, enabling users to adapt the radio to various communication scenarios and environments.

## Safety Precautions and Best Practices

Operating Baofeng radios involves more than just communication—it requires a responsible approach to ensure smooth functionality, respect for regulations, and consideration for other users. By adhering to safety precautions and best practices, users can maximize the potential of their radios while contributing to a conducive communication environment.

### Regulatory Compliance

- **Local Regulations:** Different regions have specific regulations governing radio frequency usage, transmission power limits, and permissible frequencies. Users must familiarize themselves with local regulations to avoid interference issues and ensure lawful operation.
- **Licensing Requirements:** Some frequencies or transmission modes might require users to obtain appropriate radio licenses or authorizations. Compliance with these requirements is crucial to operate within legal boundaries.

### Operating Environment Considerations

- **Interference Mitigation:** Be mindful of potential interference that could disrupt communication. Avoid transmitting on frequencies that might interfere with other users and switch frequencies if interference is encountered.
- **Responsible Spectrum Usage:** Conserve the radio spectrum by refraining from unnecessary or prolonged transmissions. Efficient communication involves concise and clear messaging without unnecessary chatter.
- **Environmental Conditions:** Understand the impact of environmental factors on signal transmission. Factors like terrain, weather, and obstacles can affect communication range and clarity. Adjust settings or frequencies accordingly to optimize communication under varying conditions.

### Proper Radio Usage

1. **Radio Etiquette:** Follow established radio etiquette guidelines. Wait for an appropriate pause before transmitting, avoid interrupting ongoing communications, and maintain professionalism in speech.

2. **Channel Selection:** Choose the appropriate channel or frequency for your communication needs. Use designated frequencies and channels for specific purposes to avoid interference and maintain communication integrity.

3. **Avoiding Unauthorized Modifications:** Refrain from unauthorized modifications to the radio's hardware or software. Unauthorized changes can lead to malfunctions, violation of regulations, or reduced device lifespan.

## Safety and Health Considerations

1. **Battery Safety:** Follow recommended charging practices outlined in the manual to ensure battery longevity and safety. Overcharging or improper handling can affect the battery's performance and safety.

2. **Radio Heat Dissipation:** During prolonged use, radios might generate heat. Ensure proper ventilation and avoid covering or obstructing the device, allowing heat to dissipate naturally.

3. **User Health:** Maintain an appropriate distance between the radio and your body to minimize potential electromagnetic exposure. Use accessories like earpieces or headsets to reduce direct exposure during extended usage.

## Communication Practices

1. **Effective Communication:** Clearly convey messages without unnecessary repetition or prolonged transmissions. Ensure your message is concise, to the point, and free of unnecessary background noise.

2. **Monitoring and Listening:** Practice active monitoring of channels before transmitting. Listening to ongoing communications helps avoid interrupting or overlapping transmissions.

3. **Emergency Communication Protocols:** Understand and adhere to emergency communication protocols in case of emergencies. Keep channels clear for emergency communications and follow established procedures.

## Radio Maintenance and Handling

1. **Regular Maintenance:** Periodically inspect and maintain your Baofeng radio. Check for loose connections, clean the device, and ensure all components are functioning correctly.

2. **Proper Handling:** Handle the radio with care to avoid physical damage. Protect it from impact, moisture, extreme temperatures, and other potentially damaging elements.

**Training and Knowledge**

1. **User Education:** Educate yourself and other users on radio operation, regulations, and best practices. Proper training ensures responsible and efficient radio usage.

2. **Continuous Learning:** Stay updated on new regulations, technological advancements, and best practices. Engage in forums, attend workshops, or seek guidance from experienced users to enhance your knowledge.

## Popular Baofeng Models

Baofeng, renowned for its diverse range of radios catering to myriad applications, stands as a premier choice among global radio users. Tailoring models for amateur radio operators, outdoor adventurers, security personnel, and diverse user groups, Baofeng's offerings vary in form factor, features, and compatibility, empowering users to select models aligned with their specific needs and preferences.

Understanding the nomenclature of Baofeng radio models is crucial, providing insightful cues into their fundamental characteristics. Baofeng typically designates model names with alphanumeric codes, like the iconic "UV-5R." Deciphering this code reveals essential traits:

- **"UV":** Signifying a dual-band model capable of transmitting on both UHF (Ultra High Frequency) and VHF (Very High Frequency) bands, these radios demonstrate versatility across a broad spectrum of frequencies. The distinction between UHF and VHF bands lies in their propagation characteristics. UHF frequencies excel in urban and indoor settings, while VHF frequencies offer superior outdoor coverage. Baofeng radios, such as the UV-5R, encompass both bands, enabling users to select the most suitable frequency for diverse communication needs.

- **"5":** The numeric segment denotes the radio's series or generation. Baofeng employs sequential numbering, elevating the digit to differentiate versions or iterations of the same model. Each iteration may showcase enhancements, updates, or alterations in design or functionality. A higher numeric value implies a more recent or advanced version of the radio. For instance, a hypothetical "UV-6R" might emerge as a subsequent version of the "UV-5R," integrating improvements or modifications.

- **"R":** As the final character, "R" often signifies "Repeater." It indicates the radio's capability to utilize repeaters, extending communication ranges significantly. Repeaters serve as devices that receive and retransmit signals, effectively amplifying a radio signal's reach. This feature substantially enhances a radio's effective communication range. Additionally, various suffixes such as "W" or "X" within model names might denote specific variations or unique features. For example, "W" could designate a waterproof model, while "X" might identify an advanced or specialized edition with distinctive functionalities.

Now, let's delve into popular Baofeng models and their exceptional features:

*Baofeng BF-F8HP*

The Baofeng BF-F8HP stands out as a highly regarded model renowned for its heightened power and superior performance when compared to the UV-5R. This radio has garnered a dedicated following among amateur radio operators and outdoor enthusiasts seeking extended range and enhanced capabilities. Let's delve into its key features that set it apart:

1. **High Power Output:** The "HP" designation in the model name stands for "High Power," and the BF-F8HP lives up to this distinction. Offering an impressive power output of up to 8 watts, significantly higher than the UV-5R's standard 4 watts, this amplified power translates into an expanded communication range. Such increased power makes it ideal for long-distance communication needs.

2. **Extended Battery Life:** Equipped with an upgraded 2000mAh battery, the BF-F8HP extends its operational time significantly. This proves particularly advantageous during outdoor adventures, emergency situations, or prolonged usage scenarios where access to a charging source might be limited.

3. **Tri-Power Options:** The BF-F8HP offers three distinct power output options: High (8W), Medium (4W), and Low (1W). This tri-power feature grants users the flexibility to choose the appropriate power setting tailored to their specific communication requirements. Users can balance between the desired range and battery life based on their preferences.

4. **Dual PTT Buttons:** This model incorporates dual Push-to-Talk (PTT) buttons, facilitating seamless switching between two frequencies or communication modes. Such a feature enhances multitasking and significantly streamlines communication dynamics, particularly during group activities or situations requiring rapid frequency changes.

5. **Larger Antenna:** The BF-F8HP boasts a longer and more efficient antenna, a crucial component contributing to an extended range and improved signal quality. This enhanced antenna proves highly beneficial for outdoor enthusiasts and individuals in need of expanded communication coverage.

6. **Advanced Programming:** Compatibility with the CHIRP programming software empowers BF-F8HP users to effortlessly program the radio. This compatibility not only simplifies the programming process but also allows users to configure the radio precisely to their specifications, thereby enhancing customization possibilities.

*Baofeng BF-888S*

Known for its simplicity and cost-effectiveness, the BF-888S caters to users seeking basic yet reliable communication, boasting essential features and practical attributes:

1. **Single-Band Operation:** Operating within the UHF range, the BF-888S functions on a single band. Tailored for users requiring basic communication needs, it sidesteps the complexities associated with dual-band radios, focusing on simplicity and efficiency in its operations.

2. **Compact and Simple Design:** Boasting a compact and straightforward design, the BF-888S prioritizes user-friendly functionality. Its minimalistic interface ensures ease of use, making it an ideal choice for individuals valuing simplicity in their communication devices.

3. **Long Battery Life:** A highlight of the BF-888S is its exceptional battery life. Renowned for its energy efficiency, this radio model allows users to operate for extended periods without the hassle of frequent recharging. This feature proves particularly advantageous in scenarios where access to charging sources might be limited or sporadic.

4. **Solid Build Quality:** Despite its simplicity, the BF-888S is built to withstand the rigors of daily use. Its robust construction guarantees durability, rendering it suitable for a spectrum of professional applications and outdoor activities, ensuring reliability even in demanding environments.

5. **License-Free Operation:** Operating within the Family Radio Service (FRS) frequencies in the United States, the BF-888S eliminates the need for an amateur radio license. This accessibility makes it an appealing choice for users seeking a radio that can be used without regulatory constraints, simplifying its adoption for various communication needs.

*Baofeng UV-82*

The Baofeng UV-82 stands as a prominent dual-band radio choice, celebrated for its distinctive design and tailored features designed to cater to a broad spectrum of users. With its unique appearance and robust performance, the UV-82 has garnered a dedicated following. Let's delve into its key features that set it apart in the realm of two-way radios:

1. **Rugged Build:** The UV-82 boasts a robust and durable construction, rendering it well-suited for outdoor adventures and challenging environments. Its resilience against water and dust makes it a reliable choice for outdoor enthusiasts, hikers, and campers seeking durability in their communication gear.

2. **Dual Push-to-Talk (PTT):** Similar to the BF-F8HP, the UV-82 integrates dual Push-to-Talk buttons, simplifying communication between different channels or distinct groups. This feature significantly enhances efficiency and facilitates real-time communication, enabling swift transitions between communication modes.

3. **Large Battery Capacity:** Equipped with a substantial 2800mAh battery, the UV-82 ensures extended operating time, delivering reliable power for prolonged durations. Whether navigating the wilderness or engaging in emergency response situations, the high-capacity battery provides users with uninterrupted communication.

4. **LED Flashlight:** The UV-82 model includes a built-in LED flashlight, serving as a practical and potentially life-saving tool during power outages or when navigating in the dark. This feature adds an extra layer of utility to the radio, catering to a wide array of applications.

5. **FM Broadcast Reception:** Offering users access to FM radio broadcasts, the UV-82's built-in FM receiver provides not only entertainment but also crucial access to information during emergencies. This versatile feature adds significant value to the radio's utility.

6. **Compatibility:** The UV-82 demonstrates compatibility with various Baofeng accessories and third-party add-ons, allowing users to personalize their radio setup according to their specific needs. This adaptability extends to external microphones, earpieces, and antennas, providing users with enhanced customization options.

*Baofeng DM-5R*

The DM-5R, a digital dual-band radio with advanced features, stands out in the realm of digital radio technology, embracing versatility and forward-looking capabilities, with noteworthy features:

1. **Digital and Analog Operation:** A standout feature of the DM-5R is its support for both digital and analog modes. This versatility allows users the flexibility to transition seamlessly from analog to digital communication while ensuring compatibility with existing analog

systems. This transition capability is particularly beneficial for users navigating the shift from traditional analog to more modern digital communication methods.

2. **Dual-Time Slot:** With the ability to manage dual-time slots, the DM-5R enables two simultaneous conversations or data streams on a single frequency. This efficient utilization of channel resources maximizes communication efficiency, allowing users to optimize channel usage effectively.

3. **Color Display:** Sporting a vibrant color display, the DM-5R presents an aesthetically pleasing and user-friendly interface. The clarity of the color screen enhances user experience by providing easily accessible settings and vital information, ensuring effortless navigation through menus.

4. **Text Messaging:** The DM-5R facilitates text messaging, enabling users to send and receive messages directly through the radio. This feature proves invaluable in scenarios where discreet communication is necessary or in environments where voice communication might not be feasible.

5. **Enhanced Audio Quality:** Leveraging digital audio processing, the DM-5R delivers superior audio quality compared to traditional analog models. The digital technology ensures clear and crisp sound transmission, contributing to enhanced communication effectiveness across various environments.

6. **Built-In GPS:** An outstanding feature of the DM-5R is its integrated GPS functionality. This built-in GPS capability enables users to utilize location tracking and sharing, proving highly beneficial for outdoor activities, navigation, and emergency response situations where precise location information is vital.

7. **Promiscuous Mode:** Supporting promiscuous mode, the DM-5R allows users to monitor multiple talk groups on the same frequency. This feature significantly enhances situational awareness, proving especially useful for professionals coordinating activities across different groups by providing comprehensive oversight.

*Baofeng GT-3*

Recognized for its versatility and user-friendly design, the GT-3 caters to amateur radio operators and outdoor enthusiasts, incorporating user-friendly features and standout attributes:

1. **Tri-Power Options:** Aligning with the BF-F8HP model, the GT-3 offers users three distinct power output options: High (8W), Medium (4W), and Low (1W). This trifecta of

power settings empowers users to fine-tune their communication needs, striking the ideal balance between range and battery life based on specific requirements.

2. **Dual PTT Buttons:** The GT-3 incorporates the convenience of dual Push-to-Talk (PTT) buttons. This functionality simplifies communication across two different channels or with separate groups, streamlining multitasking and facilitating smoother communication, especially during group activities where swift switching between channels is essential.

3. **Rugged Build:** Crafted with a robust and durable construction, the GT-3 is tailored for outdoor ventures and challenging environments. Its sturdy design ensures resilience against water and dust, guaranteeing consistent and reliable performance even in adverse conditions, making it an ideal companion for outdoor enthusiasts in various situations.

4. **Color Display:** Sporting a vibrant color display, the GT-3 offers users a clear and visually appealing interface. The vivid color screen enhances user experience by providing easy access to settings, monitoring critical information, and navigating through menus with heightened visibility and clarity.

5. **Compatibility:** Similar to several Baofeng models, the GT-3 boasts compatibility with a diverse range of Baofeng accessories and third-party add-ons. This flexibility empowers users to personalize their radio setup by connecting external microphones, earpieces, antennas, and other accessories, tailoring their equipment to specific preferences and operational needs.

*Baofeng UV-5R - The Iconic Choice*

The Baofeng UV-5R stands tall as an emblematic choice amongst Baofeng radios. Its allure is attributed to several distinct features and applications that resonate with a diverse user base.

The UV-5R offers dual-band functionality, enabling access to both VHF and UHF frequencies. This unique capability allows users to communicate across a broad spectrum of frequencies, making it highly adaptable to various communication needs.

For amateur radio enthusiasts, often referred to as 'hams,' the UV-5R is particularly popular due to its compatibility with the 2-meter (144-148 MHz) and 70-centimeter (420-450 MHz) bands. These bands are frequently used by ham radio operators for local and regional communication. The UV-5R simplifies the process of saving and scanning through numerous channels, making it a favored choice for ham radio novices and seasoned operators alike.

The selectable power levels of the UV-5R offer users the flexibility to choose from different power settings, including low, medium, and high. This adaptability allows users to manage energy consumption while optimizing communication over varying distances, a feature highly appreciated by those in diverse operating environments.

Moreover, the radio's intuitive display and keypad facilitate easy navigation through menus and programming, ensuring a seamless user experience even for beginners. The backlit display ensures visibility in low-light conditions, making it a dependable tool for emergency responders and outdoor enthusiasts alike.

CTCSS (Continuous Tone-Coded Squelch System) and DCS (Digital-Coded Squelch) are integral features of the UV-5R, providing users with sub-audible tones that enable access to specific repeaters or communication channels without interference. This enhances privacy and reduces unwanted chatter on shared frequencies, critical in scenarios requiring confidential or focused communication.

The UV-5R's capability for dual-watch and dual-receive allows users to monitor two channels simultaneously or receive on two channels consecutively, optimizing efficiency and multitasking during communication.

Beyond its core functionalities, the UV-5R encompasses additional features such as an FM radio receiver. This feature allows users to access weather alerts and entertainment during downtime, a crucial aspect in emergency preparedness and outdoor activities.

The versatility of the UV-5R extends to its battery options, providing users with a rechargeable lithium-ion battery for extended operating time. Additionally, the radio is adaptable to alkaline batteries when a rechargeable source is unavailable, ensuring its reliability in remote or emergency situations.

Baofeng UV-5R's array of accessories and upgrades further contributes to its appeal. Baofeng offers a wide range of accessories such as external microphones, extended-capacity batteries, antennas, and programming cables. These accessories allow users to customize and enhance the radio according to their specific requirements, contributing to its adaptability and versatility.

The UV-5R's scanning and memory capabilities streamline access to frequently used frequencies, simplifying the communication process and enabling quick access to channels of interest.

The inclusion of the Voice-Operated Transmit (VOX) function in the UV-5R enables hands-free communication, a valuable asset in situations where manual keying of the microphone is impractical or inconvenient.

Squelch levels can be adjusted on the UV-5R, allowing users to filter out background noise and improve communication clarity, especially in noisy environments or adverse conditions.

Basic encryption options are available on the UV-5R, ensuring private conversations for users requiring secure communication, adding an extra layer of privacy and confidentiality.

The UV-5R's sturdy build quality ensures its resilience in adverse conditions, making it suitable for outdoor use and demanding environments. Its durability against rough handling and exposure to various weather conditions solidifies its reliability.

**The UV-5R's Popularity and Reasons Behind It**

The Baofeng UV-5R's immense popularity can be attributed to several compelling reasons:

1. **Affordability:** The UV-5R's accessibility due to its affordability makes it an attractive option, particularly for amateur radio operators, outdoor enthusiasts, and individuals with budget constraints.

2. **Versatility:** The UV-5R's dual-band operation, compatibility with diverse accessories, and adaptability to various applications make it a versatile choice catering to a wide range of user needs.

3. **Ease of Use:** Its user-friendly design ensures that both novice and experienced users can operate it effectively without requiring extensive technical knowledge.

4. **Support for Emergency Communication:** The UV-5R's broad coverage, robust construction, and FM radio functionality make it a dependable tool in emergency situations, allowing users to stay informed and coordinate effectively.

5. **Community and Resources:** The UV-5R benefits from a large and active user community, resulting in abundant online resources such as user guides, programming software, and forums. This wealth of resources simplifies the learning curve for users and aids in maximizing the radio's capabilities.

6. **Upgradability:** The ability to enhance and personalize the UV-5R through a range of accessories ensures that users can tailor the radio to meet their specific requirements, contributing to its widespread appeal.

7. **Strong Reception:** The UV-5R's sensitivity and reception capabilities, even in challenging radio environments, are frequently praised by users, affirming its effectiveness across diverse scenarios.

8. **Global Availability:** Baofeng radios, including the UV-5R model, are accessible globally, ensuring that users from different countries can utilize the same model, fostering a widespread user base and community.

## The Evolution of Baofeng UV-5R

Over time, the Baofeng UV-5R has undergone several iterations and improvements, leading to enhanced models such as the UV-5R V2+, UV-5R Plus, and the UV-5R III. These variations introduced subtle changes and enhancements in design, power, and features while retaining the core functionality and affordability that made the UV-5R famous.

## Applications of Baofeng UV-5R

The Baofeng UV-5R finds utility in various applications across different user groups and scenarios:

1. **Amateur Radio (Ham Radio):** Ham radio operators widely employ the UV-5R for local and regional communication across different frequency bands. Its affordability and ease of use make it a favored choice, especially among newcomers to the hobby.

2. **Emergency Preparedness:** The UV-5R's broad frequency coverage, dual-band capability, and FM radio function make it an essential tool in emergency preparedness kits and disaster response plans, ensuring reliable communication during crises.

3. **Outdoor Activities:** Outdoor enthusiasts, including hikers, campers, and hunters, rely on the UV-5R for communication in remote areas where cellular coverage may be unreliable or nonexistent. Its durability and versatility make it an indispensable companion during outdoor adventures.

4. **Security and Surveillance:** Security professionals and private security firms utilize the UV-5R for on-site communication, coordination among teams, and monitoring, benefiting from its ruggedness and reliable performance.

5. **Construction and Trades:** Construction crews, maintenance teams, and tradespeople use the UV-5R for on-site communication, enabling seamless coordination and enhancing safety in workplaces.

6. **Recreational Use:** Radio enthusiasts, hobbyists, and individuals interested in exploring two-way radio communication utilize the UV-5R for recreational purposes, experimenting with different frequencies and settings.

**Search and Rescue Missions:** Search and rescue teams find the UV-5R invaluable due to its portability, versatility, and ability to function in challenging environments, aiding in effective coordination during rescue operations.

# *Chapter 2:* **Understanding Radio Fundamentals**

Radio communication serves as a cornerstone for modern communication, enabling the wireless transmission of information across vast distances. This chapter aims to delve into the foundational aspects of radio communication, elucidating fundamental concepts and terminologies essential for comprehending this intricate yet pivotal domain.

## What is Radio Communication?

Radio communication is a technology that allows the transmission and reception of information using radio waves. It has been a significant development in the field of wireless communication, connecting people across long distances. Radio communication has played a crucial role in various industries and has had a profound impact on society, enabling the exchange of information, entertainment, and coordination of essential services.

Radio waves are a form of electromagnetic radiation that travels at the speed of light. They have the ability to propagate through the atmosphere and other materials, making long-distance communication possible without the need for physical connections. By harnessing radio waves, radio communication has revolutionized the way people communicate, providing wireless connectivity and enabling real-time information exchange.

### Historical Development

The history of radio communication can be traced back to the late 19th century when inventors made significant contributions to the understanding and application of radio waves. Guglielmo Marconi, Nikola Tesla, and Heinrich Hertz were among the pioneers in this field.

Marconi's work on wireless telegraphy in the late 1890s marked a significant milestone in the development of practical radio communication systems. He successfully demonstrated the transmission and reception of telegraph signals over long distances without the need for physical wires. Marconi's experiments and advancements in wireless telegraphy laid the foundation for further developments in radio communication.

Around the same time, Nikola Tesla conducted experiments on wireless power transmission and made important discoveries related to radio waves. Tesla's work on wireless technology and his visionary ideas contributed to the advancement of radio communication.

Heinrich Hertz's experiments and theoretical work on electromagnetic waves provided fundamental insights into the properties and behavior of radio waves. Hertz's experiments confirmed the existence of radio waves and their ability to be generated, transmitted, and detected.

## Basic Principles of Radio Communication

Radio communication relies on the principles of modulation and demodulation. Modulation involves varying the characteristics of a carrier wave in accordance with the information to be transmitted. The carrier wave is a high-frequency electromagnetic wave that acts as a carrier for the information.

There are various modulation techniques used in radio communication, including amplitude modulation (AM), frequency modulation (FM), and phase modulation (PM). In AM, the amplitude of the carrier wave is varied in proportion to the information signal. FM involves varying the frequency of the carrier wave based on the information signal, while PM varies the phase of the carrier wave.

The transmission process in radio communication involves the modulation of the carrier wave with the desired information signal. The modulated signal is then sent through an antenna, which radiates the signal into space. On the receiving end, another antenna captures the radiated signal, and the demodulation process extracts the original information from the modulated carrier wave.

## Components of a Radio Communication System

A typical radio communication system consists of several components that work together to facilitate the transmission and reception of signals.

The transmitter is responsible for processing the information to be transmitted and modulating it onto the carrier wave. It consists of various electronic circuits, such as amplifiers, oscillators, and modulators, that shape and manipulate the signal before it is transmitted.

The receiver captures the modulated signal using an antenna and then demodulates it to recover the original information. The receiver contains circuits for signal amplification, filtering, and demodulation.

Antennas play a crucial role in radio communication as they are responsible for the transmission and reception of radio waves. The antenna on the transmitter side radiates the modulated signal into space, while the antenna on the receiver side captures the radiated signal for further processing.

The medium through which the radio waves propagate is an essential component of radio communication. The medium can be the atmosphere, space, or other materials like buildings and obstacles. The properties of the medium can affect signal propagation, leading to phenomena such as reflection, diffraction, and attenuation.

## Applications of Radio Communication

Radio communication finds widespread application across diverse sectors:

### 1. Telecommunications

Radio communication serves as the backbone of telecommunications systems, enabling voice and data transmission across long distances. This application encompasses a wide array of technologies, including:

- **Two-way Radios**: Utilized in industries such as construction, security, hospitality, and public services for instant communication between individuals or teams within a limited range.
- **Cellular Networks**: Mobile phones rely on radio waves to connect with cellular towers, allowing users to communicate over vast areas via network infrastructure.
- **Satellite Communications**: Satellite phones and communication systems leverage satellites to transmit signals globally, ensuring connectivity in remote or inaccessible regions.

### 2. Broadcasting

Radio broadcasting remains a powerful medium for disseminating information, entertainment, news, and music to the public. AM (Amplitude Modulation) and FM (Frequency Modulation) radio stations cater to diverse audience preferences, offering a wide range of content and reaching a broad spectrum of listeners.

### 3. Aviation and Maritime Industries

In aviation, radios are vital for communication between aircraft, air traffic control towers, and ground personnel. Similarly, in maritime industries, radios facilitate essential communication between ships, ports, maritime authorities, and coast guards, ensuring safety and efficient operations at sea.

4. **Public Safety and Emergency Services**

Police departments, fire brigades, emergency medical services, and disaster response teams heavily rely on radio communication. These services use radios for swift and effective coordination during emergencies, enabling quick response, critical information sharing, and operational management.

5. **Amateur Radio**

Amateur radio enthusiasts engage in experimentation, communication activities, and public service events. They contribute to emergency communication networks, participate in contests, conduct scientific experiments, and promote international goodwill through communication exchanges.

## Advantages of Radio Communication

The significance of radio communication is underscored by its numerous advantages:

1. **Wireless Connectivity**: One of the primary advantages of radio communication is its ability to transmit information wirelessly over extensive distances. This feature eliminates the need for physical connections, offering unparalleled mobility and flexibility.

2. **Adaptability and Versatility**: Radio technology is adaptable and versatile, catering to diverse communication needs across various industries and applications. It allows for various transmission formats, frequencies, and modulation techniques, making it suitable for a wide range of scenarios.

3. **Scalability and Coverage**: Radio communication systems can be easily scaled up to cover extensive geographic areas, making them ideal for wide-area coverage in applications ranging from local networks to global satellite systems. This scalability ensures connectivity across vast regions.

4. **Reliability in Adverse Conditions**: Radio communication often remains reliable even in adverse conditions where other communication methods might face challenges. Its resilience in harsh environments, natural disasters, or emergencies makes it a critical lifeline, providing essential connectivity when needed most.

5. **Cost-Effective Solutions**: Compared to establishing wired communication infrastructure, radio communication systems often offer more cost-effective solutions. They require minimal infrastructure, making them an efficient choice for quickly deploying communication networks, especially in remote or temporary setups.

## Fundamental Concepts

Radio communication is a complex field built upon several fundamental concepts that form the backbone of wireless transmission systems. Understanding these concepts is crucial for engineers, technicians, and enthusiasts in developing, maintaining, and improving communication technologies. Here, we'll delve into the core principles that define radio communication.

### Electromagnetic Waves

Electromagnetic waves are the foundation of radio communication. These waves consist of perpendicular oscillating electric and magnetic fields that propagate through space. According to Maxwell's equations, changing electric fields produce magnetic fields, and vice versa, resulting in a self-propagating wave that moves at the speed of light. The wave properties, including frequency, wavelength, amplitude, and propagation direction, define the characteristics of electromagnetic radiation.

The propagation of electromagnetic waves occurs through various mediums, including air, vacuum, and other materials. Understanding the behavior of these waves helps engineers design antennas and communication systems that effectively transmit and receive signals over short and long distances.

### Transmitter and Receiver

A transmitter converts information into electromagnetic signals suitable for transmission. It modulates the carrier signal by superimposing the information signal onto it, using modulation techniques such as AM, FM, or PM. The transmitter amplifies the modulated signal and sends it through an antenna into the surrounding space.

The receiver captures these signals using its antenna and demodulates the received signals to extract the original information. The demodulation process reverses the modulation applied at the transmitter end, allowing the receiver to recover the transmitted data accurately.

## Modulation

Modulation alters certain properties of the carrier signal to encode information. It allows the transmission of information over radio waves by modifying characteristics like amplitude, frequency, or phase of the carrier signal. Different modulation techniques offer trade-offs in terms of bandwidth efficiency, resilience to noise, and complexity of implementation.

- **Amplitude Modulation (AM):** In AM, the amplitude of the carrier wave is varied in proportion to the instantaneous amplitude of the information signal.
- **Frequency Modulation (FM):** FM alters the frequency of the carrier wave based on the instantaneous amplitude of the modulating signal.
- **Phase Modulation (PM):** PM modifies the phase of the carrier wave in response to changes in the information signal.

Modulation techniques play a crucial role in determining the efficiency and reliability of transmitted data, making them integral to various communication systems.

## Antennas

Antennas are essential components that facilitate the transmission and reception of electromagnetic waves. They convert electrical signals into radio waves for transmission and vice versa for reception. The design and configuration of antennas depend on factors such as frequency, polarization, radiation pattern, and gain.

- Dipole antennas, with their simple design, are commonly used for transmitting and receiving signals in various applications.
- Yagi-Uda antennas, characterized by multiple elements and directional properties, are often utilized for radio and television reception.
- Parabolic antennas, known for their high gain and directionality, are employed in satellite communications and long-distance point-to-point links.

The choice of antenna type and configuration impacts the range, coverage, and efficiency of a communication system.

## Radio Frequency Spectrum

The radio frequency spectrum comprises a wide range of frequencies allocated for various communication purposes. Regulatory bodies, like the International Telecommunication Union (ITU), allocate specific frequency bands to different services and applications to prevent interference and ensure efficient utilization of the spectrum.

The spectrum encompasses diverse frequency bands:

- Very Low Frequency (VLF) and Low Frequency (LF) bands are used for submarine communications and long-range navigation.
- Medium Frequency (MF) and High Frequency (HF) bands facilitate long-distance radio broadcasting and international communications.
- Very High Frequency (VHF) and Ultra High Frequency (UHF) bands are common in television broadcasting, FM radio, and mobile communication networks.
- Microwave bands cater to point-to-point communication, satellite links, and radar systems.

Efficient management and allocation of the spectrum are crucial for accommodating the increasing demand for wireless communication services.

## Propagation

Radio waves propagate through different mediums and environments, encountering various obstacles and conditions that affect their behavior. Understanding propagation characteristics is vital for designing communication systems that adapt to diverse scenarios.

- Ground Wave Propagation occurs at lower frequencies, where radio waves follow the curvature of the Earth's surface, providing coverage for short and medium distances.
- Sky Wave Propagation involves bouncing radio signals off the ionosphere, enabling long-distance communication, especially in HF bands.
- Line-of-Sight Propagation operates at higher frequencies, requiring a clear path between the transmitter and receiver without obstructions for effective signal transmission.

Factors such as terrain, atmospheric conditions, and obstacles influence how radio waves propagate, impacting the design and deployment of communication systems.

## Noise and Interference

Noise and interference pose challenges in radio communication, degrading the quality of transmitted signals. Sources of noise include thermal noise generated by electronic components, atmospheric disturbances, and external interference from other devices or signals.

Techniques such as error correction coding, shielding, frequency hopping, and spread spectrum modulation help mitigate the impact of noise and interference, ensuring reliable communication in noisy environments.

## Signal Strength and Attenuation

Signal strength measures the power of a received signal and is crucial for determining the quality and reliability of communication. As radio waves propagate through space or encounter obstacles, they experience attenuation, causing a reduction in signal strength. Understanding signal attenuation helps in designing systems that compensate for signal losses, ensuring robust communication links.

## Multiplexing

Multiplexing allows multiple signals to share a common transmission medium efficiently. Frequency Division Multiplexing (FDM) and Time Division Multiplexing (TDM) are common multiplexing techniques used in various communication systems.

- FDM divides the frequency spectrum into multiple channels, allowing simultaneous transmission of different signals in each channel.
- TDM allocates time slots within a transmission cycle to different signals, enabling sequential transmission of multiple data streams.

Multiplexing techniques optimize the use of available bandwidth, catering to multiple users or information streams within a communication system.

## Demodulation

Demodulation is the process of extracting the original information from the modulated carrier signal at the receiver end. The receiver's demodulator reverses the modulation process applied at the transmitter, recovering the transmitted data accurately.

The choice of demodulation technique depends on the modulation method used in transmission, ensuring effective signal decoding and information retrieval.

# Radio Communication Terminology

Radio communication terminology encompasses a broad range of terms and concepts crucial to understanding the intricacies of wireless communication. Here is an in-depth exploration of various terminologies used in radio communication:

## Frequency and Wavelength

- **Frequency**: The number of oscillations or cycles per second measured in Hertz (Hz). Radio frequencies span from kHz (kilohertz) to GHz (gigahertz).
- **Wavelength**: The distance between successive peaks of a wave. It is inversely proportional to frequency. Wavelength ($\lambda$) = Speed of Light (c) / Frequency (f).

## Modulation Techniques

- **Amplitude Modulation (AM)**: Modulates signal by varying the amplitude of the carrier wave. Commonly used in broadcasting.

- **Frequency Modulation (FM)**: Modulates signal by varying the frequency of the carrier wave. Provides better audio quality.
- **Phase Modulation (PM)**: Modulates signal by changing the phase of the carrier wave. Used in digital transmission.

## Transmission

- **Transceiver**: Device capable of both transmitting and receiving radio signals, common in mobile phones and two-way radios.
- **Simplex**: One-way communication; examples include walkie-talkies where only one party can speak at a time.
- **Duplex**: Allows two-way simultaneous communication—full-duplex (simultaneous transmission and reception) or half-duplex (one direction at a time).

## Components

- **Antenna**: Essential for transmitting and receiving radio waves.
- **Transmitter**: Converts electrical signals into radio waves for transmission.
- **Receiver**: Device that picks up and processes incoming radio signals.

## Signal Strength and Quality

- **Signal-to-Noise Ratio (SNR)**: Ratio of desired signal strength to background noise, indicating signal quality.
- **RSSI (Received Signal Strength Indicator)**: Measures power level received by a device's antenna.

## Protocols and Standards:

- **Frequency Band**: A specific range of frequencies allocated for particular services or applications (e.g., AM radio, Wi-Fi).
- **Modulation Scheme**: The technique used to encode information onto a carrier wave (AM, FM, PM, etc.).
- **Protocol**: Set of rules governing data transmission between devices, ensuring compatibility and reliability.

## Propagation and Interference

- **Propagation**: The way radio waves travel from the transmitter to the receiver, affected by factors like frequency, obstacles, and atmospheric conditions.
- **Interference**: Unwanted signals disrupting the transmission and reception of desired signals.

**Types of Radio Communication**

- **Analog Communication**: Uses continuous signals for transmission, like AM and FM.
- **Digital Communication**: Utilizes discrete signals (0s and 1s) for transmission, offering improved accuracy and noise immunity.

**Radio Bands and Services**

- **HF (High Frequency)**: Long-range communication, used in aviation and maritime services.
- **VHF (Very High Frequency)**: Commonly used for FM radio, television broadcasting, and air traffic control.
- **UHF (Ultra High Frequency)**: Used in GPS, cell phones, and satellite communication.

**Encryption and Security**

- **Encryption**: Process of encoding information to prevent unauthorized access.
- **Security Protocols**: Measures and standards implemented to secure wireless communication from hacking or unauthorized access.

**Multiplexing and Demultiplexing:**

- **Multiplexing**: Technique that combines multiple signals for transmission over a single channel.
- **Demultiplexing**: Process of separating combined signals back into individual components.

**Radio Wave Propagation**

- **Line-of-Sight (LOS)**: Direct path between transmitter and receiver without obstruction.
- **Ionospheric Propagation**: Refers to radio waves bouncing off the ionosphere, enabling long-distance communication.

**Noise and Interference Mitigation:**

- **Fading**: Variations in signal strength due to interference, obstacles, or atmospheric conditions.
- **Equalization**: Techniques to compensate for signal distortions caused by transmission medium or interference.

**Radio Frequency Identification (RFID):**

- **RFID Tags**: Devices using radio waves to transmit data for identification and tracking purposes.

**Spectrum Allocation and Regulations:**

- **FCC (Federal Communications Commission)**: Regulatory body responsible for managing radio frequency allocation in the United States.
- **Spectrum Allocation**: The process of assigning specific frequency bands for various communication services to avoid interference.

# *Chapter 3:* **Setting Up Your Baofeng Radio**

Setting up a Baofeng radio involves several essential steps to ensure proper functionality and optimal performance. Here's a comprehensive guide detailing the setup process:

## Initial Battery Charging

Charging the battery of your Baofeng radio is a crucial first step before using the device. Proper initial charging ensures optimal battery performance, longevity, and safe operation.

### Importance of Initial Battery Charging

- **Battery Performance** Baofeng radios typically come with rechargeable battery packs or support for standard batteries. Proper initial charging is vital as it conditions the battery and maximizes its performance over its lifespan.
- **Battery Longevity:** Correctly charging the battery initially helps in setting its capacity and maintaining its health. This process helps prevent overcharging or undercharging, which can degrade the battery's capacity over time.
- **Safety:** Following the manufacturer's guidelines for initial charging is essential for safety. Overcharging or using incompatible chargers can lead to overheating, battery damage, or, in extreme cases, pose a safety risk.

### Steps for Initial Battery Charging

1. **Battery Installation**

Ensure the Baofeng radio is turned off before inserting the battery pack or batteries. Follow these steps for proper installation:

- Open the battery compartment located at the back of the radio.
- Insert the provided rechargeable battery pack or standard batteries according to the polarity markings.
- Ensure a snug fit to prevent accidental disconnection during use.

2. **Selecting the Charger**

Baofeng radios typically come with a specific charger designed for their battery packs. It is crucial to use the provided or recommended charger to prevent damage to the battery or radio.

### 3. Charging Procedure

Follow these steps for safe and effective initial charging:

- Connect the charger to the radio or the battery pack.
- Plug the charger into a suitable power outlet.
- Observe the LED indicator on the charger; it typically changes color or turns off when the battery is fully charged.
- Avoid leaving the battery unattended while charging and ensure the charging area is well-ventilated.

### 4. Charging Duration

Manufacturers often provide guidelines for the initial charging duration. It's advisable to follow these recommendations to prevent overcharging, which can negatively impact battery life.

### 5. Checking Charge Status

Once the charging cycle is complete, verify the charge status:

- Unplug the charger from the power outlet.
- Disconnect the charger from the radio or battery pack.
- Check the LED indicator on the charger; a full charge may be indicated by a change in color or turning off.

### 6. Battery Conditioning

For optimal battery performance, consider fully charging and discharging the battery a few times initially. This process, known as battery conditioning, helps establish the battery's maximum capacity.

### 7. Monitoring Battery Temperature

While charging, ensure the battery does not become excessively hot. Elevated temperatures can indicate a problem with the charger or the battery itself. If the battery becomes unusually hot, stop charging immediately and seek assistance from the manufacturer or a qualified technician.

### 8. Follow Manufacturer's Guidelines

Always refer to the user manual or manufacturer's instructions specific to your Baofeng radio model for detailed information on initial battery charging procedures and safety precautions.

# Attaching the Antenna

Attaching the antenna to your Baofeng radio is a fundamental step in ensuring efficient transmission and reception of radio signals. The antenna plays a crucial role in capturing and emitting radio waves, impacting the device's range and overall performance.

**Importance of Attaching the Antenna**

1. **Signal Reception and Transmission**

- The antenna is the primary interface between your Baofeng radio and the surrounding electromagnetic environment.
- Proper attachment ensures optimal signal reception and transmission, enhancing communication range and clarity.

2. **Signal Quality**

- A securely attached antenna minimizes signal loss or interference, improving the overall quality of transmitted and received signals.

3. **Device Protection**

- Properly attaching the antenna protects the radio from potential damage caused by mismatched impedance or exposure to excessive RF energy.

**Steps for Attaching the Antenna**

1. **Selecting the Antenna**

- Baofeng radios typically come with a detachable antenna, often a whip-style antenna.
- Ensure you are using the appropriate antenna designed for your radio model and intended frequency range.

2. **Identifying the Antenna Port**

- Locate the designated antenna port on the top of your Baofeng radio.
- The port is usually a threaded connector designed to accommodate the antenna's screw-on mechanism.

3. **Antenna Attachment Procedure**

- Align the threaded end of the antenna with the antenna port on the radio.
- Gently insert the antenna into the port, ensuring a proper fit without applying excessive force.

- Rotate the antenna clockwise until it securely fastens onto the port.

## 4. Avoid Over-tightening

- While it's essential to secure the antenna firmly, avoid over-tightening to prevent damage to the antenna connector or port.
- Hand-tightening is generally sufficient to ensure a secure connection.

## 5. Checking Antenna Connection

- After attaching the antenna, gently try to wiggle it to ensure it's firmly attached without any loose movements.
- Verify that the antenna is stable and not easily dislodged, especially during movement or operation.

## 6. Antenna Positioning

- For optimal performance, ensure the antenna is in an upright position and not obstructed by surrounding objects.
- Keep the antenna vertical and away from the body while in use to minimize signal absorption by the human body.

## 7. Periodic Inspection

- Regularly check the antenna's condition for any signs of wear, damage, or loosening.
- Clean the antenna periodically to remove dust or debris that may accumulate on its surface.

### Considerations and Tips

- **Antenna Length and Frequency Range:** Different antennas may be suitable for specific frequency ranges. Ensure compatibility with the frequencies you intend to use.
- **Portable Antennas:** Consider using portable or external antennas for extended range or specialized applications, such as outdoor use or specific frequency bands.
- **Antenna Removal:** When not in use or for storage, it's advisable to remove the antenna from the radio to prevent accidental damage.

## Powering On and Off

Powering on and off your Baofeng radio is a foundational step in using the device effectively and safely. Understanding the proper procedures for turning the radio on and off ensures smooth operations and extends the device's lifespan.

**Steps for Powering On and Off**

    1. **Powering On the Radio**

**a. Locating the Power Button/Knob:**

- Identify the power button or knob on your Baofeng radio. It is usually marked with a power symbol or labeled as 'Power' or 'On/Off.'

**b. Turning On the Radio:**

- To turn on the radio, press and hold the power button or turn the power knob in the designated direction until the device powers on.
- Listen for an audible tone or watch for the display to illuminate, indicating that the radio is on.
- Some models might have a separate position for powering on; follow the manufacturer's instructions specific to your Baofeng radio model.

**c. Initialization and Startup:**

- Upon powering on, the radio may undergo a startup process, displaying the manufacturer's logo or initializing the settings. Allow the device to complete this process before use.

    2. **Powering Off the Radio**

**a. Initiating Shutdown:**

- To turn off the radio, press and hold the power button or turn the power knob in the designated direction until the device begins the shutdown process.
- Ensure you hold the button or knob long enough for the radio to initiate the shutdown sequence.

**b. Confirming Shutdown:**

- Observe the display or listen for audio cues indicating that the device is powering off. Some models might display a shutdown message or tone.
- Wait for the device to complete the shutdown process before proceeding further.

**c. Verifying Power-Off State:**

- Once powered off, ensure that the display, indicator lights, or any audio output from the radio have ceased to confirm that the device is fully turned off.

**Additional Considerations:**

### a. Soft Power-Off vs. Hard Power-Off:

- Some Baofeng radios might have a soft power-off function that initiates a graceful shutdown process, ensuring data integrity and proper closure of applications. Others may have a hard power-off where the device is immediately powered off without a shutdown sequence.

### b. Power-Saving Features:

- Explore the radio's settings for power-saving features, such as auto power-off timers, which can automatically turn off the device after a specified period of inactivity to conserve battery life.

**Best Practices and Tips**

- **Avoid Forceful Actions:** Always handle the power button or knob gently to prevent damage to the switch or internal components.
- **Follow Manufacturer's Instructions:** Refer to the user manual for specific instructions related to powering on and off your Baofeng radio model.
- **Periodic Reboots:** Occasionally power cycling the radio (turning it off and then on again) can help refresh its operations and resolve minor glitches.
- **Low Battery Indication:** Pay attention to low battery warnings or indicators to avoid unexpected shutdowns during critical communication.

# *Chapter 4*: **Navigating the Baofeng Radio Menu**

Navigating the menu of your Baofeng radio is fundamental to accessing various functions and settings essential for optimal performance. Understanding the display, buttons, and menu structure allows users to program channels, adjust settings, and customize their radio experience. Here's a comprehensive guide on navigating the Baofeng radio menu:

## Display and Buttons Overview

Let's delve into the details of the various buttons and their functionalities commonly found on Baofeng radios. These buttons and controls play a crucial role in operating the device and accessing its features:

### Power/Volume Knob:

- **Function:** This knob typically serves a dual purpose. It powers the radio on and off when turned and also controls the volume by rotating it.
- **Operation:** Rotating the knob clockwise turns the radio on and increases the volume. Counterclockwise rotation powers the radio off and decreases the volume.

### Numeric Keypad:

- **Function:** Comprising numbers (0-9) and sometimes additional characters or functions, this keypad allows direct input of frequencies, channels, or settings.
- **Operation:** You can input specific frequencies or channel numbers directly using these keys, facilitating quick access to desired frequencies or channels.

### Function Keys (FUNC):

- **Function:** These keys serve multiple functions based on their combinations with other buttons, enabling access to secondary or special features of the radio.
- **Operation:** Pressing these keys in combination with other buttons or during specific operations allows access to additional settings or features.

### Menu Button:

- **Function:** The menu button is central to accessing the radio's menu system, enabling users to navigate various settings and functions.

- **Operation:** Pressing this button typically enters the menu mode, allowing users to scroll through different menu options using the up/down buttons.

**Up/Down Buttons:**

- **Function:** These buttons facilitate navigation within the menu system, enabling users to scroll through various options or settings.
- **Operation:** Pressing the up button navigates upwards through menu options, while the down button navigates downwards, allowing selection of specific functions or settings.

**PTT (Push-to-Talk) Button:**

- **Function:** The PTT button is crucial for initiating transmissions. When pressed, it allows the user to transmit their voice or data.
- **Operation:** Pressing this button activates the microphone for transmitting. Releasing the button ends the transmission, allowing for reception.

**A/B or VFO/MR Button (Dual Watch):**

- **Function:** This button switches between memory mode (MR) and VFO mode (Variable Frequency Oscillator) or channel A/B for dual-watch capabilities.
- **Operation:** Toggling this button allows users to switch between stored memory channels and manually entered frequencies or channels.

**Monitor Button:**

- **Function:** The monitor button is used to open the squelch and listen to weak or distant signals or break through interference.
- **Operation:** Pressing this button temporarily disables the squelch, allowing reception of signals regardless of their strength or quality.

**Exit/CLR Button:**

- **Function:** The exit or clear button allows users to exit from a menu, cancel an operation, or clear input during programming.
- **Operation:** Pressing this button typically exits the current menu or mode, returning the user to the previous screen or function.

**Lock Button:**

- **Function:** The lock button is used to lock or unlock the keypad, preventing accidental button presses or changes to settings.
- **Operation:** Pressing this button for a few seconds activates or deactivates the keypad lock, ensuring that settings remain unchanged.

**Scan Button:**

- **Function:** This button initiates scanning for active frequencies or channels, allowing users to search for available transmissions.
- **Operation:** Pressing this button triggers the radio to scan through programmed channels or frequencies to find active signals.

**Side Buttons (Model-dependent):**

- **Function:** Some models may feature additional side buttons, such as an emergency alarm, flashlight, or programmable shortcut keys.
- **Operation:** These buttons offer quick access to specific functions or settings depending on the radio model and user preferences.

## Programming Your First Channel

Programming your first channel on a Baofeng radio involves setting specific frequencies, tones, and other parameters to create a personalized communication channel. This process allows for quick access to desired frequencies or channels for efficient communication.

### 1. Accessing Channel Programming Mode

a. **Entering the Menu**

- Press the menu button on your Baofeng radio to access the menu system.
- Use the up/down buttons to navigate through the menu options until you find the channel programming or frequency setting.

b. **Selecting Channel Memory Slot**

- Once in the programming mode, select an empty memory slot where you want to store your first channel frequency.
- Most radios offer multiple memory slots for storing different frequencies or channels.

### 2. Inputting Frequencies

a. **Frequency Entry**

- Use the numeric keypad on the radio to input the desired frequency for your channel.
- Enter the frequency digits one by one, ensuring accuracy and precision.

b. **Step Size Adjustment (if available)**

- Some radios allow adjusting the frequency step size for finer or coarser tuning.
- Set the step size to match the frequency increments you intend to use for programming.

3. **Assigning Channel Name (Optional)**

a. **Accessing Channel Naming (if available)**

- Baofeng radios might allow assigning names or labels to channels for easier identification.
- Navigate to the channel naming option in the menu system.

b. **Inputting Channel Name**

- Use the numeric keypad or navigation buttons to input the desired channel name or label.
- Follow the on-screen prompts to enter characters or symbols for naming the channel.

4. **Setting Transmit and Receive Tones (CTCSS/DCS)**

a. **Tone Selection**

- Baofeng radios often support CTCSS (Continuous Tone-Coded Squelch System) or DCS (Digital-Coded Squelch) to filter out unwanted signals.
- Access the tone settings in the menu and select the appropriate tone for transmit and receive functions, if needed.

b. **Match Transmit/Receive Tones (if required)**

- Ensure that the transmit and receive tones are set to match those of the intended communication group or channel to access transmissions effectively.

5. **Saving the Channel**

a. **Confirmation and Saving**

- Once all necessary settings are inputted, review the entered information to ensure accuracy.
- Follow the on-screen instructions to confirm and save the programmed channel into the selected memory slot.

b. **Finalization and Exit**

- After saving the channel, exit the programming mode or menu system to return to the main display or standby mode.
- Verify that the programmed channel appears correctly on the display for future access.

**Tips and Considerations**

1. **Frequency Selection:** Choose frequencies allowed for use in your region and comply with legal regulations and licensing requirements for transmitting on specific frequencies.

2. **Channel Organization:** Organize your channels or frequencies logically, grouping them based on usage (e.g., emergency channels, local repeaters, favorite contacts) for easier access.

3. **Verify Settings:** Double-check all settings, including frequencies, tones, and channel names, to ensure accuracy before saving the channel.

4. **User Manual Reference:** Refer to the user manual specific to your Baofeng radio model for detailed instructions and guidance on programming channels, as different models may have varying procedures.

# Adjusting Volume and Squelch

Adjusting volume and squelch settings on a Baofeng radio is crucial for ensuring clear and effective communication. Volume control regulates the audio output level, while squelch helps eliminate background noise when no signal is present.

1. **Adjusting Volume**

a. **Volume Control Knob/Buttons**

- Baofeng radios typically feature a volume control knob or dedicated buttons to adjust audio output.
- Rotating the knob clockwise or pressing the volume-up button increases the volume, while counterclockwise rotation or pressing the volume-down button decreases it.

b. **Optimizing Volume Level**

- Adjust the volume to a comfortable level that allows clear and audible communication without causing distortion or discomfort.
- Ensure the volume is adequate to hear incoming transmissions but not excessively loud to avoid potential hearing damage.

2. **Squelch Control**

a. **Understanding Squelch**

- Squelch is a function that mutes audio output when no signal is received or when signals are weak, eliminating background noise.
- Adjusting the squelch threshold helps filter out unwanted noise, ensuring that only strong and relevant signals are heard.

b. **Accessing Squelch Settings**

- Enter the menu system on your Baofeng radio and navigate to the squelch control or settings option.

c. **Adjusting Squelch Level**

- Baofeng radios typically use numerical values to indicate squelch levels, such as 0 to 9 or 1 to 10.
- Lower squelch levels (closer to 0 or 1) allow weaker signals to be heard but may introduce more background noise.
- Higher squelch levels (closer to 9 or 10) mute weaker signals, eliminating background noise but potentially missing weaker transmissions.

d. **Optimizing Squelch for Clarity**

- Experiment with different squelch levels while monitoring the radio for static or unwanted noise.
- Set the squelch just above the level where background noise disappears, allowing for clear reception without cutting off weaker signals.

**Best Practices for Volume and Squelch Adjustments:**

a. **Test in Different Environments**

- Adjust volume and squelch settings in various environments to account for differences in noise levels and signal strengths.
- Test settings in both quiet and noisy environments to find a balance between clarity and noise reduction.

b. **Regular Monitoring and Adjustment**

- Regularly monitor incoming transmissions and adjust volume and squelch settings as needed to maintain optimal clarity and reception.
- Revisit settings based on changing environmental conditions or communication requirements.

c. **Documenting Ideal Settings**

- If using specific channels or frequencies frequently, note down the ideal volume and squelch settings for quick reference in similar conditions.

d. **User Manual Reference**

- Consult the user manual specific to your Baofeng radio model for detailed instructions and guidance on adjusting volume and squelch settings, as procedures may vary between models.

# *Chapter 5:* **Basic Radio Operations**

Understanding basic radio operations is crucial for effective communication using your Baofeng radio. This includes transmitting and receiving signals, adjusting transmit power, utilizing dual watch and dual receive functionalities, and using CTCSS and DCS tones for improved communication. Here's an in-depth discussion of these fundamental radio operations:

## Transmitting and Receiving

Transmitting and receiving are fundamental functions of any radio communication device, including Baofeng radios. Let's delve into transmitting and receiving in detail:

### Transmitting

Transmitting refers to the process of sending out a signal from your radio to communicate with other stations. When transmitting, it is essential to understand the proper procedures and techniques to ensure effective communication. Let's delve into the key aspects of transmitting:

1. **Frequency Selection:** The first step in transmitting is selecting the desired frequency. Radios operate within specific frequency bands allocated for various purposes, such as amateur radio, commercial use, or public safety. It is important to ensure that you are operating within the legal limits of your license class and following any applicable regulations. Use the tuning controls on your radio to select the desired frequency within the allocated band.

2. **Transmit Power:** Adjusting the transmit power is crucial to achieving effective communication. The power output of a radio is typically measured in watts (W) or milliwatts (mW). Higher power levels allow for greater signal reach, but they consume more battery power and may cause interference if used irresponsibly. It is essential to understand the power limits specified by your license class and any applicable regulations. Different license classes may have different power restrictions, and certain bands may have specific power limits. Adjust the power output of your radio to an appropriate level for the communication you are initiating, using the power control settings on your radio.

3. **Microphone Usage:** To transmit your voice or message, most radios are equipped with a microphone. The microphone captures your voice and converts it into an electrical signal that is then transmitted through the radio. When transmitting, hold the microphone close

to your mouth and speak clearly and concisely. Articulate your words to ensure that your message is intelligible to other operators.

4. **Push-to-Talk (PTT) Button:** The push-to-talk (PTT) button is a feature commonly found on radios. It is a button that needs to be pressed or held while transmitting. When you want to transmit, press the PTT button to activate the transmitter and allow your voice to be transmitted over the airwaves. Release the PTT button after you have finished transmitting your message.

5. **Speaking Protocol:** When transmitting, it is important to adhere to proper speaking protocols to ensure effective communication. Speak clearly and at a moderate pace. Avoid speaking too quickly, as it can make it difficult for others to understand you. Use standard communication practices such as using phonetic alphabets, proper call signs, and clear language to convey your message accurately.

## Receiving

Receiving involves listening and decoding signals from other stations. It is equally important to understand the receiving process to effectively interpret incoming signals. Let's explore the key aspects of receiving:

1. **Frequency Tuning:** To receive signals, ensure that your radio is tuned to the desired frequency. Use the tuning controls on your radio to adjust the frequency and align it with the frequency of the transmitting station you intend to listen to. It is crucial to tune your radio accurately to receive clear signals.

2. **Volume Adjustment:** Set the volume control on your radio to an appropriate level to hear incoming signals clearly. Adjust the volume according to your environment, ensuring it is audible but not overly loud.

3. **Active Listening:** Receiving requires active listening. Pay close attention to the sounds coming from your radio's speaker. Listen for any incoming signals, including CQ calls (general calls seeking a response), conversations, or emergency transmissions. Actively monitor the frequency for any relevant communication.

4. **Note Taking:** If you hear important information or need to remember details, it is helpful to take notes. Note-taking ensures that you capture essential details accurately and helps you refer back to the information when needed.

5. **Responding:** When appropriate, respond to incoming calls or contribute to ongoing conversations. Before transmitting a response, ensure that you follow the proper communication protocol, such as waiting for a pause or using appropriate call signs. Key the microphone, hold the PTT button, and speak clearly and concisely. Release the PTT button after transmitting your response.

## Best Practices for Transmitting and Receiving

Transmitting and receiving signals effectively in radio communication requires adherence to best practices. These practices ensure clear and efficient communication while minimizing interference and maximizing signal quality. Let's explore some of the best practices for transmitting and receiving signals:

### Transmitting Best Practices

- **Adhere to Frequency and Power Regulations:** Prior to transmitting, it is crucial to familiarize yourself with the regulations and licensing requirements pertaining to your specific frequency band and power output. Different license classes and frequency bands have distinct limitations. Operating within the legal boundaries is essential to avoid penalties and prevent interference with other radio operators.
- **Master Proper Microphone Technique:** When transmitting, hold the microphone near your mouth and speak directly into it with clarity. Avoid speaking too softly or positioning the microphone too far away, as it can result in weak or distorted audio. Practicing good microphone technique ensures that your voice is effectively transmitted to other operators.
- **Communicate Clearly and Concisely:** Enunciate your words and maintain a moderate pace while speaking to ensure that your message is easily comprehensible to fellow operators. Steer clear of mumbling or speaking too rapidly, as it may impede understanding. Employ standard communication practices, such as using phonetic alphabets and employing clear language, to convey your message accurately.
- **Follow Proper Protocol:** Adhere to established communication protocols and etiquette. Wait for your turn to transmit, and avoid interrupting ongoing conversations unless necessary. Use appropriate call signs, identifiers, and standard phrases to ensure smooth and efficient communication. Following proper protocol promotes effective and respectful communication among operators.
- **Monitor Your Transmission:** While transmitting, listen to your own transmission on a separate receiver or through monitoring functions on your radio if available. This allows you to verify the quality and clarity of your transmission. Monitoring your transmission helps you identify any issues and make adjustments if needed.

**Receiving Best Practices**

- **Accurate Frequency Tuning: Ensure** that your radio is accurately tuned to the desired frequency. Use the tuning controls to align with the transmitting station's frequency. Accurate frequency tuning is crucial to receiving clear and intelligible signals. Be mindful of any frequency drift or changes in the transmitting station's frequency and adjust accordingly.

- **Active Listening:** Engage in active listening while receiving signals. Pay close attention to the sounds coming from your radio's speaker and actively monitor the frequency. Listen for incoming calls, conversations, or relevant information. Actively engaging in listening helps you stay aware of the communication happening around you.

- **Optimize Volume Control:** Adjust the volume control on your radio to an appropriate level to hear incoming signals clearly. Set the volume according to your environment, ensuring it is audible but not excessively loud. Proper volume control allows you to distinguish signals from background noise effectively.

- **Take Notes:** Taking notes can be helpful, especially when receiving important information or instructions. Note-taking ensures that you capture essential details accurately and helps you refer back to the information when needed. Develop a shorthand or notation system that works for you to take quick and concise notes.

- **Respond Appropriately:** When appropriate, respond to incoming calls or contribute to ongoing conversations. Before transmitting a response, ensure that you follow the proper communication protocol. Wait for a pause or use appropriate call signs to indicate your intent to transmit. Key the microphone, hold the PTT button, and speak clearly and concisely. Release the PTT button after transmitting your response.

**General Best Practices**

- **Maintain Proper Radio Configuration:** Ensure that your radio is configured correctly for the desired communication. Set the appropriate transmit power level, modulation type, and any other relevant settings. Regularly check and adjust your radio's configuration to match your intended communication requirements.

- **Antenna Considerations:** Choose the appropriate antenna for your communication needs. Different antennas have different characteristics and radiation patterns. Consider factors such as range, directionality, and environmental conditions when selecting an antenna. Position and orient the antenna correctly to optimize signal transmission and reception.

- **Avoid Unnecessary Interference:** Be mindful of potential interference issues while transmitting and receiving. Use the minimum power necessary to achieve the desired communication range. Avoid transmitting on frequencies or bands where you do not have authorization. Respect other operators' communication and avoid unnecessary interruptions or lengthy transmissions that may hinder effective communication.

- **Continuous Learning and Improvement:** Stay engaged in continuous learning to enhance your skills and knowledge in radio communication. Keep updated with the latest techniques, technologies, and regulations. Engage in discussions with experienced operators, participate in training programs, and seek opportunities to expand your proficiency. Regular practice and improvement contribute to becoming a more effective and confident operator.

By adhering to these best practices, you can ensure clear, efficient, and respectful communication while transmitting and receiving signals. Following proper procedures, maintaining good etiquette, and continuously improving your skills will enhance your overall radio communication experience.

## Adjusting Transmit Power

Adjusting transmit power on a Baofeng radio is a critical aspect of optimizing communication range and conserving battery life. It involves selecting the appropriate power level at which the radio transmits signals.

### What is Transmit Power?

Transmit power refers to the amount of electrical power that a radio transmitter emits when transmitting a signal. It is typically measured in watts (W) or milliwatts (mW). Transmit power determines the strength of the radio signal that is radiated from the antenna.

### Importance of Adjusting Transmit Power

Adjusting transmit power is of paramount importance in radio communication. The transmit power level directly affects the range, signal quality, and overall performance of the communication system.

### 1. Communication Range

One of the primary reasons why adjusting transmit power is vital is its direct impact on the communication range. Transmit power determines how far your signal can travel and be received by other stations. By increasing the transmit power, you can extend the range at which your signals can be detected and decoded. This is particularly important in situations where long-range communication is required, such as in emergency response operations or during communication between remote locations.

On the other hand, in scenarios where communication is intended for shorter distances, excessive transmit power may not be necessary and can lead to unnecessary interference. By adjusting the transmit power according to the specific communication range requirements, you can optimize the use of resources and minimize interference with other nearby communication systems.

## 2. Signal Quality

Transmit power plays a crucial role in ensuring clear and reliable communication by affecting the signal quality. When the transmit power is appropriately adjusted, it helps to overcome signal losses, propagation obstacles, and interference, resulting in a stronger and more robust signal at the receiving end. This, in turn, improves the overall clarity and intelligibility of the communication.

In situations where the transmit power is set too low, the received signal may become weak, leading to poor quality, increased noise, and susceptibility to interference. On the other hand, excessive transmit power can cause overdriving and distortion, leading to a degraded signal quality. By adjusting the transmit power to an optimal level, you can achieve a balance that ensures a strong and clear signal while minimizing interference and distortions.

## 3. Battery Life and Power Efficiency

Another crucial aspect of adjusting transmit power is its impact on battery life and power efficiency. Transmitting at higher power levels consumes more energy, which can significantly reduce the operational time of battery-powered communication systems. In situations where power sources are limited or not readily available, efficient utilization of power becomes crucial.

By adjusting the transmit power intelligently, you can strike a balance between achieving the desired communication range and conserving battery life. Lower transmit power levels can help extend the operational time of battery-powered radios, enabling longer communication sessions without the need for frequent recharging or replacement of batteries. This is particularly important in remote or emergency situations where access to power sources may be limited.

## 4. Interference Management

Interference is a common challenge in radio communication, especially in crowded frequency bands or densely populated areas. Adjusting transmit power is an effective way to manage and mitigate interference issues. By using the minimum power necessary to achieve the desired communication range, you can reduce the chances of causing interference to nearby communication systems operating on the same or adjacent frequencies.

Excessive transmit power can lead to overloading and saturation of receivers, resulting in interference with neighboring channels or receivers. By adjusting the transmit power to an optimal level, you can minimize the risk of interference and promote coexistence with other communication systems, ensuring smooth and uninterrupted communication for all parties involved.

### 5. Compliance with Regulations

Regulatory compliance is a critical aspect of radio communication. Different license classes and frequency bands have specific regulations and limitations regarding transmit power. It is crucial to understand and adhere to these regulations to avoid penalties and interference with other operators. By adjusting the transmit power within the prescribed limits, you ensure compliance with the regulatory requirements, promoting responsible and lawful use of the radio spectrum.

**Factors to Consider When Adjusting Transmit Power**

When adjusting transmit power, several factors need to be considered to ensure effective communication:

### 1. Regulatory Requirements

Regulatory requirements play a crucial role in determining the maximum allowable transmit power for a particular frequency band and license class. It is important to be aware of and adhere to these regulations to avoid interference with other users and potential legal consequences. Different countries and regulatory bodies have specific rules and limitations regarding transmit power, and it is essential to comply with these requirements.

### 2. Communication Distance

The desired communication distance is a key factor when adjusting transmit power. The transmit power level should be appropriate for the intended range of communication. For shorter-range communications, such as within a building or a local area, lower transmit power levels may be sufficient. Using higher transmit power in such scenarios can cause unnecessary interference and waste energy. On the other hand, for long-range communications, such as point-to-point or satellite communications, higher transmit power may be necessary to overcome signal losses and reach the desired destination. By considering the communication distance, you can adjust the transmit power accordingly to optimize the signal strength for effective communication.

### 3. Antenna Gain

Antenna gain is the measure of how effectively an antenna radiates or receives signals in a specific direction. When adjusting transmit power, it is important to consider the gain of the antenna being used. Antennas with higher gain focus the radiated power in a specific direction, resulting in increased signal strength in that direction. By using an antenna with higher gain, you can achieve the desired communication range with lower transmit power. This allows for improved power efficiency and reduced interference in other directions. Conversely, antennas with lower gain may require higher transmit power to achieve the same communication range.

### 4. Environmental Conditions

Environmental conditions have a significant impact on radio signal propagation and, consequently, on the required transmit power. Factors such as terrain, obstructions, atmospheric conditions, and interference sources can affect the signal strength and quality. In areas with dense vegetation, buildings, or other obstructions, higher transmit power may be necessary to overcome signal attenuation caused by these obstacles. In contrast, in open areas with a clear line of sight, lower transmit power levels may be sufficient for effective communication.

### 5. Interference Considerations

Interference management is crucial for maintaining clear and reliable communication. When adjusting transmit power, it is important to consider potential interference sources and minimize the chances of causing interference to other communication systems. Using excessive transmit power can lead to overdriving receivers and causing interference with nearby radios operating on the same frequency or adjacent frequencies. By adjusting the transmit power to an appropriate level, you can strike a balance between achieving the desired communication range and avoiding unnecessary interference. Consider the proximity of other radios and their operating frequencies to ensure interference-free communication.

In addition to these factors, it is important to note that adjusting transmit power alone may not always be sufficient to overcome all communication challenges. Other techniques, such as using different modulation schemes, employing error correction coding, or selecting alternative communication frequencies, may also be necessary in certain situations.

It is recommended to gather information, conduct tests, and seek advice from experienced operators or radio communication professionals to determine the optimal transmit power settings for specific communication requirements. By considering regulatory requirements, communication distance, antenna gain, environmental conditions, and interference considerations, you can adjust the transmit power effectively to achieve reliable and efficient radio communication.

## Best Practices for Adjusting Transmit Power

To optimize the transmit power and ensure effective communication, consider the following best practices:

- **Start with Lower Power:** Begin with a lower transmit power setting and gradually increase it if needed. Starting with lower power helps conserve battery life and minimizes the risk of interference. Monitor the signal quality and range and increase the transmit power incrementally until the desired communication range is achieved.
- **Test and Observe:** When adjusting transmit power, it is essential to test and observe the results. Monitor the signal strength and quality at the receiving end and gather feedback from the receiving station. Assess the impact of the transmit power adjustments on the received signal. This feedback will help you fine-tune the transmit power to optimize communication performance.
- **Consider Signal Strength Indicators:** Many radios provide signal strength indicators that display the received signal strength. Utilize these indicators to gauge the effectiveness of your transmit power adjustments. Monitor the signal strength as you change the transmit power and aim for a level that provides a strong and reliable signal without unnecessary overdriving.
- **Balance Power and Quality:** Strive to strike a balance between transmit power and signal quality. While higher transmit power may extend the range, it can also introduce interference or distortions. Optimize the transmit power to achieve a reliable signal without causing unnecessary interference to others. Consider the specific communication requirements, operating environment, and available power resources to find the optimal balance.
- **Regularly Evaluate and Adjust:** Radio communication conditions can change over time. Factors such as atmospheric conditions, terrain, and interference sources may vary. It is important to regularly evaluate the performance of your communication system and adjust the transmit power as needed. Periodically reassess the range, signal quality, and battery consumption to ensure that the transmit power is still optimized for the current conditions.
- **Consider Power Conservation:** In scenarios where power conservation is crucial, such as during emergency operations or when operating on limited power sources, adjust the transmit power to the minimum level necessary to achieve reliable communication. This helps extend battery life and ensures that power resources are used efficiently.

- **Seek Expert Advice:** If you are unsure about the appropriate transmit power settings for your specific communication requirements, seek advice from experienced operators or radio communication professionals. They can provide guidance based on their expertise and knowledge of the operating environment.

By considering the factors mentioned above and following these best practices, you can adjust the transmit power effectively to optimize communication range, signal quality, and battery consumption. Remember to comply with regulatory requirements, regularly evaluate performance, and adapt to changing conditions to ensure successful and efficient radio communication.

## Dual Watch and Dual Receive

Baofeng radios, such as the popular UV-5R series, typically have two VFOs (Variable Frequency Oscillator) that can be programmed with different frequencies. The dual watch feature enables users to monitor the activity on one VFO while maintaining communication on the other. The radio continuously checks for signals on the secondary VFO and switches to that frequency temporarily when activity is detected. Once the activity ends, the radio returns to the primary VFO. This allows users to keep track of two frequencies without missing important communications.

Baofeng radios also offer dual receive functionality, which allows users to actively receive signals on two different frequencies simultaneously. This means that users can actively transmit and communicate on one frequency while monitoring or receiving signals on another frequency without interrupting their ongoing communication. Dual receive provides enhanced situational awareness and the ability to respond promptly to incoming transmissions.

### Applications of Dual Watch and Dual Receive in Baofeng Radios

Baofeng radios with dual watch and dual receive capabilities find applications in a wide range of scenarios:

1. **Public Safety and Emergency Services:** Dual Watch and Dual Receive are crucial in public safety and emergency service operations. They enable simultaneous monitoring of primary communication channels and secondary emergency or tactical frequencies. This allows emergency responders, such as police, fire, and medical personnel, to stay connected with their teams while receiving critical updates from dispatch or emergency channels. It ensures that emergency services can efficiently coordinate their efforts, respond to incidents, and quickly adapt to changing situations.

2. **Event Management and Security:** In event management and security scenarios, Dual Watch facilitates seamless communication and coordination. Event organizers can assign a primary frequency for event management operations and a secondary frequency for security or emergency communications. This enables real-time updates and immediate responses to security incidents or emergencies. Dual Watch ensures that event staff can monitor both channels simultaneously, allowing for efficient management of crowd control, incident response, and overall event safety.

3. **Search and Rescue Operations:** Dual Watch is invaluable during search and rescue operations. Search teams can monitor the primary frequency for team communications while simultaneously listening to a secondary frequency dedicated to emergency distress calls or coordination with other rescue teams. This enables effective communication among search team members and ensures that critical emergency calls or distress signals are not missed. Dual Watch enhances the overall efficiency and safety of search and rescue missions.

4. **Outdoor Activities and Adventure Sports:** Dual Watch is highly beneficial for outdoor activities and adventure sports, such as hiking, camping, mountaineering, and boating. Users can program a primary frequency for group communication and a secondary frequency for weather updates or emergency broadcasts. This allows outdoor enthusiasts to stay connected with their group members while remaining informed about changing weather conditions or any emergency alerts in the area. Dual Watch enhances safety and ensures efficient communication during outdoor adventures.

5. **Amateur Radio Operations:** Dual Watch is widely used by amateur radio operators to monitor multiple frequencies or bands. It allows operators to listen to a calling frequency while simultaneously monitoring a second frequency for ongoing conversations or important announcements within the amateur radio community. This feature enhances the overall amateur radio experience, promotes engagement, and facilitates efficient communication among operators.

6. **Construction and Industrial Sites:** In construction and industrial settings, Dual Watch supports effective communication and coordination among teams. Different workgroups or departments can be assigned separate frequencies for their respective operations, while a secondary frequency is designated for site-wide announcements or emergency communications. Dual Watch enables workers to monitor their primary

frequency while staying aware of crucial site-wide updates or emergency alerts, ensuring efficient and safe operations.

7. **Security and Surveillance:** Dual Watch is valuable in security and surveillance applications. Security personnel can monitor a primary frequency for team communications and a secondary frequency for surveillance or monitoring channels. This allows them to stay connected with their team members while simultaneously gathering information from surveillance sources. Dual Watch enhances situational awareness and enables prompt responses to security incidents or suspicious activities.

8. **Community and Neighborhood Watch:** Dual Watch is beneficial in community and neighborhood watch programs. Participants can monitor a primary frequency for community-wide communication and a secondary frequency for local law enforcement or emergency services channels. This enables real-time updates from both community members and official channels, improving cooperation and response to incidents. Dual Watch enhances community safety and fosters effective communication within neighborhood watch groups.

9. **Volunteer Organizations and Nonprofits:** Dual Watch is commonly used by volunteer organizations and nonprofits involved in disaster response, community service, or public events. These organizations can assign a primary frequency for their internal communication and a secondary frequency for coordinating with other organizations or receiving updates from official channels. Dual Watch ensures that volunteers stay connected with their team while actively participating in broader coordination efforts.

10. **Educational Institutions and Campus Security:** Dual Watch finds application in educational institutions and campus security operations. Campus security personnel can monitor a primary frequency for their internal communications and a secondary frequency for law enforcement or emergency services channels. This allows for seamless communication and coordination between campus security and external emergency responders. Dual Watch ensures quick responses to security incidents and enhances overall campus safety.

**Activation of Dual Watch and Dual Receive**

To activate the dual watch and dual receive features in Baofeng radios, follow these general steps:

1. **Power on the Radio:** Turn on the Baofeng radio by pressing the power button.

2. **Access VFO Mode:** Switch the radio to VFO (Variable Frequency Oscillator) mode. This mode allows you to manually select and configure the frequencies.

3. **Select Primary Frequency:** Choose the primary frequency on which you want to maintain continuous communication. This frequency will be the main channel you transmit and receive on.

4. **Set Secondary Frequency:** Select the secondary frequency that you want to monitor or receive signals from. This frequency will be the channel you switch to temporarily when activity is detected.

5. **Enable Dual Watch:** Activate the dual watch feature in the radio's settings or menu. The process for enabling dual watch may vary depending on the specific Baofeng radio model. Refer to the user manual for detailed instructions on accessing and configuring this feature.

6. **Set Dual Watch Parameters:** Configure the dual watch parameters, such as the dwell time or scan delay. The dwell time determines how long the radio stays on the secondary frequency when activity is detected before switching back to the primary frequency. Adjust the dwell time based on your preferences and operational requirements.

7. **Save Settings:** Once you have configured the dual watch settings, save the changes in the radio's memory. This ensures that the settings are retained even when you power off the radio.

## Tips for Effective Usage

To make the most of the dual watch and dual receive features, consider the following tips:

- **Configure Frequencies:** Ensure that you program the desired primary and secondary frequencies into the VFOs of your Baofeng radio. This allows you to switch between the frequencies seamlessly.
- **Set Dwell Time Appropriately:** Adjust the dwell time or scan delay according to your operational needs. A shorter dwell time allows for quicker switching between frequencies, but you may risk missing parts of conversations. A longer dwell time ensures that you capture more of the activity on the secondary frequency, but it may delay your return to the primary frequency.
- **Prioritize Channels:** If you are monitoring multiple channels, prioritize them based on their importance and relevance to your operations. This ensures that you pay more attention to critical channels and respond promptly to their activity.

- **Practice and Familiarize:** Spend time practicing with the dual watch and dual receive features before using them in critical situations. Familiarize yourself with the radio's controls, menus, and settings to ensure smooth operation and quick response when needed.
- **Stay Attentive:** While using dual watch or dual receive, it is important to remain attentive to both frequencies. Be prepared to switch frequencies when activity is detected and actively listen to theincoming transmissions. Avoid getting distracted or solely focusing on one frequency, as it may cause you to miss important information on the other.
- **Regularly Update Frequencies:** Keep your frequencies up to date and relevant to your communication needs. Add or remove frequencies as required, ensuring that you are monitoring the most appropriate channels for your specific situation.
- **Respect Radio Etiquette:** Adhere to proper radio etiquette and protocols when using dual watch or dual receive features. Wait for a break in conversations before transmitting, and avoid unnecessary interruptions or prolonged transmissions that may interfere with others' communications.

## Using CTCSS and DCS Tones

CTCSS (Continuous Tone-Coded Squelch System) and DCS (Digital-Coded Squelch) are methods used in two-way radios like Baofeng radios to enable selective communication and reduce interference in radio transmissions. These technologies play a crucial role in enhancing communication reliability, especially in scenarios where multiple users share the same frequency channel.

### Understanding CTCSS (Continuous Tone-Coded Squelch System)

CTCSS is an analog encoding method that operates by adding a low-frequency audio tone to the transmitted signal. This tone, typically between 67 and 250.3 Hz, is inaudible to human ears. Each tone represents a specific code or "sub-audible tone," which acts as a signal key to open the receiving radio's squelch only when it detects the matching tone. This technique filters out unwanted transmissions on the same frequency that don't have the corresponding tone, reducing interference from other users.

### Working Principle of CTCSS

- When a radio transmits a signal with a specific CTCSS tone, only radios tuned to the same frequency and with the matching CTCSS code will open their squelch and receive the transmission.
- It helps prevent receiving unwanted signals from other users who might be transmitting on the same frequency but using different CTCSS codes.

## Advantages of CTCSS

- **Reduced Interference:** By using different CTCSS tones, users on the same frequency can communicate without interfering with one another.
- **Enhanced Privacy:** It provides a basic level of privacy since only radios set to the same CTCSS tone can receive the transmission.
- **Selective Communication:** Allows multiple groups of users to share the same frequency without interrupting each other.

## Exploring DCS (Digital-Coded Squelch)

DCS, also known as CDCSS (Continuous Digital-Coded Squelch System), is a more advanced and secure method of encoding compared to CTCSS. Unlike CTCSS, which uses analog tones, DCS utilizes digital encoding to create a unique digital code for each channel. DCS encodes the transmitted signal with a unique digital pattern, ensuring greater reliability and resistance to false openings of the squelch.

## Working Principle of DCS

- DCS employs a digital code, usually consisting of a combination of binary bits, to encode and decode signals.
- Similar to CTCSS, the receiving radio needs to be set to the same DCS code as the transmitting radio to open its squelch and receive the signal.

## Advantages of DCS

- **Higher Security:** DCS offers increased security as it's less prone to false openings or unauthorized access compared to CTCSS due to its digital nature.
- **Improved Reliability:** Digital coding provides more precise and accurate decoding, reducing the chances of interference from signals with similar analog tones.

## CTCSS vs. DCS: A Comparison

1. **Technology:** CTCSS uses analog tones, while DCS relies on digital coding.

2. **Security:** DCS offers higher security and reliability due to digital coding, making it less susceptible to interference and false openings compared to CTCSS.

3. **Complexity:** DCS tends to be more complex to implement and might require more advanced equipment compared to CTCSS.

4. **Applications:** While CTCSS is widespread and suitable for general communication needs, DCS is employed in more critical and secure environments.

**Programming CTCSS and DCS Tones in Baofeng Radios**

Baofeng radios offer programmable options for CTCSS and DCS tones, allowing users to configure these tones for individual channels or groups. The specific steps for programming CTCSS and DCS tones may vary slightly depending on the model of the Baofeng radio, but the general process involves accessing the radio's menu settings and navigating to the CTCSS/DCS programming section. Here is a general overview of how to program CTCSS and DCS tones in Baofeng radios:

1. **Enter Frequency Mode:** Turn on your Baofeng radio and press the VFO/MR button to enter the frequency mode. This mode allows you to select and program specific channels.

2. **Select the Channel:** Use the Up/Down arrow keys or the numeric keypad to select the desired channel where you want to program the CTCSS or DCS tone.

3. **Enter Menu Mode:** Press the Menu button on your Baofeng radio to enter the menu mode. The LCD screen will display the menu options.

4. **Navigate to T-CTCS or T-DCS Settings:** Use the Up/Down arrow keys to scroll through the menu options until you find the T-CTCS (CTCSS) or T-DCS (DCS) setting. Press the Menu button to select it.

5. **Enable CTCSS or DCS**: In the T-CTCS or T-DCS settings menu, you will find options to enable or disable CTCSS or DCS tones. Select the option to enable the desired tone type (CTCSS or DCS).

6. **Set the Tone Frequency or Code:** Once you have enabled CTCSS or DCS, navigate to the tone frequency or code setting. Use the Up/Down arrow keys or the numeric keypad to enter the specific frequency or code for the desired tone. Refer to the tone chart or frequency list for the correct values.

7. **Save the Settings:** After entering the tone frequency or code, press the Menu button to save the settings. The Baofeng radio will store the programmed CTCSS or DCS tone for the selected channel.

8. **Repeat for Other Channels:** If you want to program CTCSS or DCS tones for multiple channels, repeat the above steps for each channel individually. Remember to select the desired channel before entering the menu mode and programming the tones.

9. **Test the Tones:** To ensure that the CTCSS or DCS tones are programmed correctly, you can perform a test. Set two Baofeng radios to the same channel and enable the programmed

tones on both radios. When one radio transmits, the other radio should only receive the signal if it has the correct tone programmed.

It's essential to note that the steps provided above are general guidelines and may vary depending on the specific model of your Baofeng radio. The menu options and navigation methods can differ slightly between different Baofeng radio models. Consult the user manual or the manufacturer's documentation for your specific radio model for more precise instructions.

Additionally, keep in mind that the CTCSS and DCS tones must match between radios for successful communication. Ensure that all radios involved in a communication group are programmed with the same CTCSS or DCS tones to avoid interference or miscommunication.

**Practical Applications of CTCSS and DCS Tones**

CTCSS and DCS tones have various practical applications in two-way radio communications. Here are some common scenarios where the use of CTCSS and DCS tones in Baofeng radios can be beneficial:

1. **Public Safety and Emergency Services:** CTCSS and DCS tones are extensively used by public safety agencies and emergency services, including police departments, fire departments, and paramedic services. These tones allow different units or teams to communicate securely and privately on shared radio channels. Each unit can be assigned a unique CTCSS or DCS tone, ensuring that only radios programmed with the corresponding tone can receive their transmissions. This helps prevent unauthorized access to sensitive information and reduces interference from other users.

2. **Security and Surveillance:** CTCSS and DCS tones play a crucial role in security and surveillance applications. Security personnel, such as private security firms, retail security teams, and event security staff, use two-way radios with CTCSS or DCS tone capabilities to maintain secure and private communication. By programming specific tones, security teams can establish private channels and prevent unauthorized individuals from listening to their transmissions. This enhances the confidentiality of sensitive security-related information and ensures effective coordination among team members.

3. **Hospitality and Event Management:** CTCSS and DCS tones find practical applications in the hospitality industry and event management. Hotels, resorts, and large event venues often use two-way radios for internal communication among staff members. By assigning different CTCSS or DCS tones to various departments or teams, such as housekeeping,

maintenance, and front desk, each group can communicate privately without interference from other departments. This helps streamline operations, improve response times, and maintain a professional and organized environment.

4. **Construction and Industrial Sites:** CTCSS and DCS tones are valuable tools on construction sites, industrial facilities, and manufacturing plants. These environments often involve multiple teams and contractors working simultaneously, requiring effective communication while minimizing interference. By using CTCSS or DCS tones, different workgroups, such as supervisors, electricians, and equipment operators, can communicate without disruption. This streamlines coordination, enhances safety, and improves overall productivity in these demanding work environments.

5. **Outdoor Recreation and Sports:** CTCSS and DCS tones are beneficial for outdoor recreation enthusiasts, such as hikers, campers, and outdoor sports enthusiasts. Radios equipped with CTCSS or DCS tones allow individuals or groups to communicate privately in remote areas, where multiple users may be operating on the same frequency. This enhances privacy, reduces interference from other outdoor adventurers, and ensures clear and uninterrupted communication, improving safety and coordination during outdoor activities.

6. **Amateur Radio (Ham Radio):** Amateur radio operators, also known as ham radio operators, utilize CTCSS and DCS tones to enhance communication on shared frequencies. Ham radio repeaters, which are stations that receive and retransmit signals, often employ CTCSS or DCS tones as access codes. By transmitting the correct tone along with their signal, amateur radio operators can activate the repeater and extend the range of their communication. This feature helps ensure that only authorized users can access the repeater system.

7. **Transportation and Logistics:** CTCSS and DCS tones play a vital role in transportation and logistics sectors. Trucking companies, delivery services, and taxi dispatch centers use two-way radios with CTCSS or DCS tones to facilitate efficient and secure communication between drivers, dispatchers, and central control. By assigning different tones to specific channels or groups, confidential information, such as delivery instructions or sensitive customer details, can be transmitted securely, reducing the risk of interception or unauthorized access.

8. **Amateur Radio Clubs and Events:** Amateur radio clubs and events, such as field days or contest operations, often utilize CTCSS and DCS tones to manage communication among participants. These tones allow organizers to create private channels for specific activities, competitions, or discussions. By using different tones for each group or function, participants can communicate within their designated channels without interference from other participants or nearby amateur radio operators.

## Best Practices for Using CTCSS and DCS Tones

To ensure effective use of CTCSS and DCS tones in Baofeng radios, consider the following best practices:

- **Standardize Tone Selection:** When using CTCSS or DCS tones in a group or organization, it is important to standardize the tone selection across all radios. This ensures compatibility and allows seamless communication between team members.
- **Test and Verify:** Before relying on CTCSS or DCS tones in critical situations, it is essential to test and verify the functionality. Conduct range tests and ensure that all radios are correctly programmed with the appropriate tones.
- **Monitor Multiple Tones:** In situations where multiple groups are using different CTCSS or DCS tones on the same frequency, it may be necessary to monitor multiple tones simultaneously. Some Baofeng radios offer the option to monitor multiple tones, allowing users to receive transmissions from multiple groups.
- **Regularly Update Programming:** As communication needs change or new groups are formed, regularly update the CTCSS and DCS tone programming in Baofeng radios. This ensures that radios are configured correctly and prevents unintentional interference or communication barriers.
- **Educate Users:** Properly educate all users on the functionality and importance of CTCSS and DCS tones. Ensure that they understand how to program and use the tones effectively. This helps maintain clear and secure communication within the organization.
- **Consider Compatibility:** When communicating with radios from other manufacturers, ensure compatibility of CTCSS and DCS tones. Different radio models may have varying tone frequency ranges or coding methods. Verify compatibility to ensure seamless communication between different radio systems.

# Chapter 6: **Programming Your Baofeng Radio**

Programming a Baofeng radio involves configuring frequencies, channels, settings, and features to suit your communication needs. To do this effectively, you'll need programming software, cables, and a basic understanding of radio operations. This chapter will cover programming software, cables, basic programming steps, storing channels, setting up scan lists, and using CHIRP software with Baofeng radios.

## Programming Software and Cables

*Programming Software*

The programming software designed for Baofeng radios plays a pivotal role in customizing and fine-tuning the settings, frequencies, and channels within the radio's memory. These software applications are specifically tailored to interface with Baofeng radios, providing users with a comprehensive platform to manage various aspects of radio communication.

### Importance and Functions

The importance of programming software lies in its ability to simplify the otherwise complex task of configuring radio settings. It serves multiple essential functions, including:

1. **Channel Configuration:** Programming software enables users to input, modify, and organize frequencies and settings for individual channels. This includes setting channel names, defining squelch levels, adjusting power output, and configuring other parameters.

2. **Bulk Editing Capabilities:** Users can efficiently perform bulk operations, such as copying settings between channels, importing/exporting channel lists in various formats, and making widespread changes across multiple channels simultaneously.

3. **Scan List Management:** These software tools facilitate the creation and customization of scan lists. Users can define specific channels or frequencies for the radio to scan through, allowing for personalized scanning sequences and preferences.

4. **User-Friendly Interface:** Most programming software offers an intuitive and user-friendly interface. The interface simplifies navigation, making it accessible even for beginners, and provides clear options for modifying settings.

5. **Compatibility:** These software applications are tailored to work seamlessly with specific Baofeng radio models. This ensures a reliable connection and efficient data transfer between the radio and the computer.

**Popular Software Options:**

When it comes to programming Baofeng radios, there are several popular software options available that can simplify the process and offer advanced features. These software programs provide a user-friendly interface for programming frequencies, channels, CTCSS/DCS tones, and other settings. Let's explore some of the popular software options for Baofeng radios.

1. **CHIRP:** CHIRP is one of the most widely used and highly regarded software programs for programming Baofeng radios. It is an open-source, cross-platform application that supports a wide range of radio models, including various Baofeng handheld radios. CHIRP offers a user-friendly interface with intuitive controls for programming frequencies, channels, tones, and other settings. It allows you to import and export data from CSV files, clone settings between radios, and manage large memory banks efficiently. CHIRP is compatible with Windows, macOS, and Linux operating systems, making it accessible to a broad user base.

2. **Baofeng UV-5R Programming Software:** Baofeng itself provides programming software specifically designed for the popular Baofeng UV-5R series radios. This software allows you to program frequencies, channels, CTCSS/DCS tones, power levels, and other settings. It features a straightforward interface with menu-driven options and supports importing and exporting data from CSV files. The Baofeng UV-5R programming software is compatible with Windows operating systems.

3. **BaoFeng VIP Programming Software:** BaoFeng VIP Programming Software is another official programming software offered by Baofeng. It supports a range of Baofeng radio models, including the UV-82, BF-F8HP, and GT-3TP. The software provides a user-friendly interface for programming channels, frequencies, tones, power levels, and other settings. It allows you to create and manage multiple configurations, clone settings between radios, and import/export data using CSV files. BaoFeng VIP Programming Software is compatible with Windows operating systems.

4. **RT Systems Programming Software:** RT Systems offers programming software specifically developed for Baofeng radios, including the UV-5R, UV-82, and BF-F8HP models. Their software provides an intuitive interface with easy-to-use controls for

programming channels, frequencies, tones, and other settings. It supports advanced features such as memory bank management, channel naming, and importing/exporting data from CSV files. RT Systems Programming Software is available for both Windows and macOS operating systems.

5. **Baofeng BF-888S Programming Software:** Baofeng BF-888S is a popular model widely used for commercial and personal communication. Specific programming software is available for this radio, allowing you to program frequencies, channels, CTCSS/DCS tones, and other settings. The software provides a straightforward interface for managing the radio's configuration efficiently. It is compatible with Windows operating systems.

6. **KG-UV Commander:** KG-UV Commander is a comprehensive programming software developed for Baofeng KG-UV series radios. It supports models such as KG-UV6D, KG-UV8D, KG-UV899, and more. The software offers a feature-rich interface with extensive options for programming channels, frequencies, tones, and other settings. It includes advanced features like memory bank management, cloning capabilities, and importing/exporting data from CSV files. KG-UV Commander is compatible with Windows operating systems.

7. **Baofeng UV-3R Programmer:** Baofeng UV-3R Programmer is tailored specifically for the Baofeng UV-3R series radios. This software provides an intuitive interface for programming frequencies, channels, CTCSS/DCS tones, and other settings. It supports importing and exporting data from CSV files, making it easy to manage large memory banks. Baofeng UV-3R Programmer is compatible with Windows operating systems.

## Advantages of Using Programming Software

Using programming software for Baofeng radios offers several advantages that enhance the programming and management experience. Let's explore the advantages in detail.

1. **Simplified Programming:** Programming software provides a simplified interface for programming Baofeng radios. It eliminates the need to navigate complex menu systems on the radio itself, making it easier to configure frequencies, channels, tones, and other settings. The software typically presents these settings in a clear and organized manner, allowing users to modify parameters with ease. This simplification reduces the learning curve and makes programming Baofeng radios more accessible to users, even those with limited technical knowledge.

2. **Efficient Configuration:** Programming software enables more efficient configuration of Baofeng radios compared to manual programming methods. It allows users to perform bulk configuration, where multiple settings can be programmed simultaneously across various channels or radios. This saves a significant amount of time, especially when dealing with a large number of radios or complex configurations. Additionally, the software often includes features like copy-paste, cloning, and importing/exporting settings, further streamlining the programming process and reducing the time required for repetitive tasks.

3. **Advanced Features and Functionality:** Programming software for Baofeng radios often provides advanced features and functionality that go beyond the capabilities of manual programming. These features can include managing memory banks, setting up scan lists, configuring power levels, enabling dual watch, and more. The software offers a comprehensive set of options that allow users to fully utilize the capabilities of their Baofeng radios. This access to advanced features enhances the functionality and versatility of the radios, expanding their usability in various scenarios.

4. **Error Reduction:** Manual programming is prone to human errors, especially when dealing with complex settings or programming multiple radios. Programming software minimizes the risk of errors by providing a controlled environment for configuration. The software ensures that parameters are entered correctly and consistently, reducing the chance of misconfiguration or programming mistakes. This not only saves time but also helps avoid potential issues that may arise from incorrect settings. By reducing errors, programming software improves the reliability and performance of Baofeng radios.

5. **Centralized Management:** Programming software allows for centralized management of multiple Baofeng radios. This is particularly beneficial in situations where numerous radios need to be programmed with similar settings or configurations. The software enables users to create templates or profiles that can be applied to multiple radios simultaneously, ensuring consistency and reducing the effort required for individual programming. Centralized management simplifies device deployment, updates, and maintenance, making it an efficient solution for organizations or individuals managing a fleet of Baofeng radios.

6. **Firmware Updates and Upgrades:** Many programming software tools provide the ability to update or upgrade the firmware of Baofeng radios. Firmware updates often introduce bug fixes, performance improvements, or new features. With programming software, users can easily install these updates without the need for complex manual

procedures. The software typically guides users through the firmware update process, ensuring a smooth and error-free experience. This allows Baofeng radio users to keep their devices up to date with the latest firmware versions, enhancing functionality and security.

7. **Collaboration and Sharing:** Programming software for Baofeng radios often facilitates collaboration and sharing among users. This is particularly useful in team environments or when working on projects with multiple contributors. Users can share their programming files or configurations, allowing others to replicate settings easily. Collaboration features streamline teamwork, promote knowledge sharing, and ensure consistency across multiple users or radios. It becomes easier to distribute standardized configurations or update settings across a fleet of Baofeng radios.

8. **Flexibility and Adaptability:** Programming software offers flexibility and adaptability in managing Baofeng radios. It allows users to quickly and easily modify configurations, adapting to changing requirements or environments. The software provides a platform for experimentation, enabling users to test different settings or scenarios without permanently altering the radio's configuration. This flexibility empowers users to optimize radio performance, customize functionality, and adapt to specific use cases or applications.

## Compatibility and Usage

Ensuring compatibility between the programming software and the Baofeng radio model is crucial. The software needs to support the specific radio model to establish proper communication and data transfer. Compatibility issues may result in failed communication between the radio and the computer, hindering the programming process.

When using programming software, it's essential to follow manufacturer instructions and guidelines provided with the software. This includes correctly connecting the radio to the computer using the programming cable, selecting the appropriate settings within the software, and following best practices for data transfer and storage.

## Advanced Customization Features

Programming software offers advanced customization options that empower users to tailor their Baofeng radios according to specific requirements:

- **Frequency Allocation:** Users can allocate specific frequencies to channels, including local repeaters, emergency frequencies, and regional communication channels, ensuring easy access and organized communication.

- **Squelch Settings:** Fine-tuning squelch levels helps in filtering out unwanted noise and ensuring clear reception, especially in areas with varying signal strengths or interference.
- **Power Output Adjustment:** Users can adjust the radio's power output settings, conserving battery life in low-power scenarios or boosting transmission power for better coverage.
- **Tone Settings:** Configuring CTCSS or DCS tones for individual channels enhances privacy and reduces interference from other users sharing the same frequency.
- **Naming and Grouping Channels:** Assigning names and grouping channels based on usage, location, or function simplifies navigation and quick access during communication.

## Efficiency in Various User Scenarios

Programming software caters to different user groups and scenarios by offering tailored solutions:

- **Amateur Radio Enthusiasts:** For amateur radio operators, the software allows precise frequency allocation for different bands and modes, facilitating seamless communication during various ham radio activities.
- **Emergency Services and Public Safety:** Software customization helps emergency service personnel program dedicated emergency frequencies, ensuring quick access during critical situations.
- **Business and Commercial Use:** Businesses can program specific channels for internal communication, allocate frequencies for security, and manage group channels for efficient team communication.
- **Recreational and Outdoor Activities:** Individuals engaging in outdoor activities or adventure sports can customize their radios for easy access to weather frequencies, park services, or group communication channels.

## Future Developments and Integration

As technology evolves, programming software is likely to integrate additional features such as:

- **Enhanced Security Measures:** Integration of encryption protocols or advanced security settings to ensure secure communication channels.
- **Cloud-Based Configurations:** Possibilities for cloud-based storage and synchronization of radio configurations, allowing easy access and management across devices.
- **Mobile Integration:** Potential for mobile applications enabling programming and configuration directly from smartphones or tablets, increasing accessibility and convenience.

*Programming Cables*

Programming cables are essential accessories for Baofeng radios, enabling users to connect their radios to a computer for programming and configuration purposes. These cables establish a direct communication link between the radio and the computer, allowing users to utilize programming software to program frequencies, channels, tones, and other settings.

## Types of Programming Cables

Baofeng radios typically use specific types of programming cables designed specifically for their models. Here, we will explore the types of programming cables commonly used for Baofeng radios.

### 1. Baofeng USB Programming Cables

Baofeng USB programming cables are the most common and widely used cables for programming Baofeng radios. They feature a USB connector on one end and a radio-specific connector on the other end. The radio-specific connector plugs into the programming port of the Baofeng radio, while the USB connector connects to the computer's USB port. These cables facilitate data transfer and communication between the radio and a computer for programming and configuration purposes.

Baofeng USB programming cables are designed to work with specific Baofeng radio models such as the UV-5R, UV-82, BF-F8HP, and many others. They are compatible with various programming software applications available for Baofeng radios. These cables offer plug-and-play functionality and are typically recognized by the computer as a virtual COM port, enabling seamless communication between the radio and programming software.

### 2. Baofeng Serial Programming Cables

Serial programming cables are an alternative to USB programming cables for Baofeng radios. They feature a serial connector on one end and a radio-specific connector on the other end. Serial programming cables were more commonly used in the past when computers had serial ports. However, many modern computers lack serial ports, so a USB-to-serial adapter is often required to connect the serial programming cable to the computer's USB port.

Baofeng serial programming cables are compatible with specific Baofeng radio models that support serial communication. These cables allow users to establish a direct connection between the radio's programming port and the computer's serial port or USB port (via an adapter). Serial programming cables are less commonly used nowadays due to the widespread availability of USB ports and the convenience of USB programming cables.

73

### 3. Baofeng Programming Adapter Cables

Baofeng programming adapter cables are specialized cables that provide compatibility between different types of programming cables and Baofeng radios. These cables feature different connectors on each end, allowing users to connect a programming cable with one type of connector to a Baofeng radio with a different type of connector.

For example, if you have a USB programming cable with a radio-specific USB connector, but your Baofeng radio has a different programming port, such as a serial port, you can use a Baofeng programming adapter cable to bridge the connection. These adapter cables are particularly useful when you have a programming cable that is not directly compatible with your Baofeng radio model.

Baofeng programming adapter cables come in various configurations, such as USB-to-serial adapters or adapters with different radio-specific connectors. They enable users to overcome compatibility issues and ensure a proper connection between the programming cable and the Baofeng radio.

## Functionality and Advantages of Programming Cables

Programming cables offer several important functions and advantages when it comes to programming Baofeng radios. Let's explore them:

1. **Data Transfer:** The primary function of programming cables is to facilitate data transfer between Baofeng radios and computers. They enable the exchange of programming data, such as frequencies, channels, tones, and other settings, between the radio and programming software running on the computer. This data transfer is crucial for efficiently programming and configuring Baofeng radios, as it eliminates the need for manual input on the radio itself.

2. **Programming Software Compatibility:** Programming cables ensure compatibility between Baofeng radios and programming software. The cables are designed to work seamlessly with specific programming software, allowing users to take advantage of the software's features and functionality for programming their radios. The software communicates with the radio through the programming cable, enabling users to configure various settings and parameters easily.

3. **Efficient Configuration:** Programming cables significantly enhance the efficiency of configuring Baofeng radios. Compared to manual programming using the radio's controls,

using programming software with a cable allows for faster and more streamlined configuration. Software interfaces provide a user-friendly environment for programming, presenting settings and options in a clear and organized manner. This simplifies the configuration process, reduces the chance of errors, and saves time, especially when programming multiple radios or complex settings.

4. **Firmware Updates and Upgrades:** In addition to programming, many programming cables also facilitate firmware updates and upgrades for Baofeng radios. Firmware updates often introduce bug fixes, performance improvements, and new features. By connecting the radio to a computer via a programming cable, users can easily install these updates using dedicated software. This ensures that Baofeng radios stay up to date with the latest firmware versions, enhancing functionality and addressing any issues or vulnerabilities.

5. **Cloning and Data Sharing:** Programming cables enable cloning and data sharing between Baofeng radios. Cloning allows users to copy the programming data from one radio to another, making it convenient to replicate settings across multiple radios. This is particularly useful for organizations or individuals managing a fleet of radios that require consistent programming. Additionally, programming cables facilitate data sharing by allowing users to save and share programming files or configurations with others. This promotes collaboration, knowledge sharing, and ensures consistent setups across different Baofeng radios.

## Choosing the Right Programming Cable

When selecting a programming cable for Baofeng radios, there are a few factors to consider:

1. **Compatibility:** Ensure that the programming cable you choose is compatible with your specific Baofeng radio model. Different Baofeng radio models may have different programming ports and connectors, so it's essential to select a cable that matches your radio's requirements. Verify the compatibility information provided by the cable manufacturer or supplier to ensure a proper fit.

2. **Cable Length:** Consider the length of the programming cable based on your needs. Longer cables provide more flexibility in terms of positioning your radio and computer. However, excessively long cables can lead to signal degradation or interference. Choose a cable length that allows for comfortable and convenient use without compromising signal quality.

3. **Cable Quality:** Opt for programming cables that are of good quality and durability. Well-constructed cables with sturdy connectors and robust wiring will ensure reliable and long-lasting performance. Read reviews or seek recommendations from other Baofeng radio users to identify reputable brands or suppliers that offer reliable programming cables.

4. **USB or Serial:** Consider the type of programming cable that best suits your computer's connectivity options. If your computer has availableUSB ports, it is recommended to choose a USB programming cable since it offers greater compatibility with modern computers. However, if your computer has a serial port or if you prefer to use a serial-to-USB adapter, a serial programming cable can still be a viable option.

5. **Software Compatibility:** Ensure that the programming cable you choose is compatible with the programming software you intend to use. Different programming cables may require specific drivers or software to establish the connection between the radio and the computer. Verify that the cable you select is supported by the programming software you plan to utilize.

## Basic Programming Steps

Programming Baofeng radios involves a systematic process to configure frequencies, channels, and settings. Understanding the basic programming steps is essential for users to effectively set up their radios for optimal communication. Let's explore the fundamental steps involved in programming Baofeng radios.

### Preparation

- **Gather Required Tools:** Ensure you have the necessary tools, including the Baofeng radio, a compatible programming cable, a computer with the appropriate programming software installed (e.g., CHIRP), and access to the desired frequencies or channels you wish to program.
- **Install Software and Drivers:** If using third-party programming software, install it on your computer. Additionally, if the programming cable requires drivers, ensure they are installed correctly to enable communication between the radio and the computer.

### Connect the Radio to the Computer

- Use the programming cable to establish a connection between the Baofeng radio and the computer. Plug one end of the cable into the radio's programming port and the other end into an available USB port on the computer.

**Read Data from the Radio**

- Open the programming software (e.g., CHIRP) on the computer and select the option to read data from the radio. This action retrieves the existing settings, frequencies, and channels stored in the radio's memory.

**Modify or Add Channels**

- Once the data is read, you can modify existing channels or add new ones. Input the desired frequencies, assign channel names, set offsets, adjust transmit power levels, and configure other parameters based on your communication needs.
- It's essential to have accurate frequency information for the channels you intend to program. This could include local repeater frequencies, emergency channels, or frequencies used by specific groups or organizations.

**Organize and Save Channel Settings**

- Organize the programmed channels in a logical sequence, such as grouping similar channels together (e.g., emergency channels, local repeaters, special event channels).
- Double-check all programmed channels for accuracy before proceeding. Ensure each channel is correctly labeled and configured with the appropriate settings.

**Write Data to the Radio**

- After making the necessary modifications, write the updated data back to the radio using the programming software. This action stores the programmed frequencies, channels, and settings into the radio's memory.
- Pay attention to prompts or confirmations from the software to ensure the data is successfully transferred to the radio without errors.

**Verify Programming and Functionality**

- Disconnect the programming cable from the radio and power on the Baofeng radio.
- Navigate through the programmed channels to verify that the frequencies, names, and settings are correctly saved and operational. Test the radio's transmission and reception capabilities on programmed channels to ensure functionality.

**Backup and Documentation**

- Consider creating a backup file of the programmed settings on your computer using the programming software. This backup can be valuable in case the radio needs resetting or if you wish to program other radios similarly.
- Document the programmed channels, frequencies, and any specific configurations for future reference or in case you need to reprogram the radio at a later time.

### Additional Tips

- **Use Channel Labels:** Assign clear and descriptive names to programmed channels for easy identification during use.
- **Scan List Configuration:** Set up scan lists in the programming software if you want the radio to scan through specific channels or frequencies.

## Storing and Managing Channels

Storing and managing channels in Baofeng radios involves organizing and storing frequencies, labels, and settings for effective communication. This process allows users to access specific channels swiftly, ensuring efficient communication in various scenarios.

### Importance of Storing and Managing Channels

### 1. Efficiency in Communication

Efficient communication is pivotal in various scenarios, from emergencies to routine operations. In emergency situations, having well-organized channels ensures quick access to crucial frequencies, enabling rapid and effective communication among response teams, first responders, and relevant authorities. It ensures that vital information is relayed promptly, contributing to swift decision-making and coordinated actions.

For everyday use, optimized channels enable users to access specific frequencies swiftly, improving communication efficiency in various settings, such as businesses, public safety departments, or community groups.

### 2. Optimized Usage

Organizing channels optimizes radio usage by categorizing frequencies based on their relevance or purpose. This categorization allows users to navigate through a range of channels efficiently. For instance, grouping channels by function (e.g., emergency, local news, public services) or geographic location (e.g., city zones, rural areas) streamlines access to specific communication needs.

An optimized channel list ensures that users can easily locate and select the necessary channels, reducing the time spent searching for frequencies and enhancing overall communication efficiency.

### 3. Ease of Navigation

Well-managed channels significantly simplify the radio's interface navigation. Clear labeling, organized groups, and logically arranged channels streamline the process of finding and selecting frequencies. This ease of navigation is vital in situations where quick access to the right channel is critical, such as during emergencies or when handling multiple communication tasks simultaneously.

Clear and intuitive organization reduces the complexity of operating the radio, allowing users to focus more on effective communication rather than navigating through a cluttered or disorganized channel list.

### 4. Reduced Confusion

A well-organized channel layout minimizes confusion and errors during radio operation. Clearly labeled and logically arranged channels help prevent accidental transmissions on incorrect frequencies. This is particularly crucial in situations where precise communication is essential, such as emergency responses or coordinated group activities.

Reducing the chances of errors in selecting channels enhances the reliability and accuracy of communication, fostering more effective interactions among users.

**Methods for Storing and Managing Channels:**

**Channel Organization**

- **Grouping by Function:** Organizing channels by function ensures that related frequencies are grouped together, facilitating quick access. For instance, grouping emergency services, public services, or specific event-related channels together.
- **Geographical Segmentation:** Categorizing channels based on geographic locations or coverage areas allows users to access local communication channels efficiently. This segmentation assists users in selecting relevant frequencies based on their geographical location.
- **Priority Channels:** Prioritizing critical channels by placing them at the beginning of the channel list ensures rapid access during emergencies or urgent situations. This arrangement ensures that crucial information is readily available without scrolling through a long list of channels.
- **Simplex and Repeater Separation:** Clearly distinguishing between simplex and repeater frequencies helps prevent accidental transmissions on the wrong type of channel. This separation avoids confusion and ensures proper communication setup.

## Labeling and Naming

- **Descriptive Channel Names:** Assigning clear and descriptive names to channels enhances their identification. For example, labeling channels with specific service names like "Fire Department Dispatch" or "Police Emergency."
- **Numeric Labeling:** Using numeric labels for channels allows for organized sorting, making high-priority or frequently used channels easily accessible. Numeric labels enable users to quickly identify and access important channels by their assigned numbers.

## Programming Software for Organization

- **Enhanced Organization Capabilities:** Programming software such as CHIRP provides advanced organizational capabilities. Users can create, modify, and arrange channel lists on a computer before transferring them to the radio. This software offers a user-friendly interface for efficient channel management and organization.

Utilizing these methods ensures a well-organized and easily accessible channel list, enabling users to navigate through frequencies swiftly and accurately during various communication needs.

## Operational Impact of Well-Managed Channels:

### Emergency Situations

- **Swift Response:** Organized channels facilitate rapid access to emergency frequencies, enabling quick and effective responses. Quick access to specific channels during emergencies, like natural disasters or accidents, allows for immediate communication, aiding in timely responses and coordinated actions.
- **Clear Communication:** Well-managed channels ensure clear and precise communication during emergencies. Clear and organized channel labels help maintain effective communication, ensuring that crucial information is conveyed accurately, preventing misunderstandings or delays.

### Public Events or Gatherings

- **Efficient Coordination:** Organized channels aid in efficiently coordinating activities during public events or gatherings. Event organizers, security teams, or volunteer groups can use designated channels for smooth communication, minimizing confusion and enhancing coordination.
- **Prevention of Congestion:** Segregated channels prevent frequency congestion during crowded events. By allocating specific channels for different groups or functions, communication remains smooth and interference-free, ensuring effective coordination.

**Everyday Use and Convenience**

- **Ease of Access:** Well-managed channels provide easy access to frequently used frequencies in daily radio operations. Whether in professional settings or recreational activities, organized channels allow users to swiftly access relevant communication channels without delays or errors.
- **Error Reduction:** Properly organized channels reduce the likelihood of errors or accidental transmissions on incorrect frequencies. This minimizes communication disruptions and ensures smoother and more reliable communication in everyday use.

# Setting Up Scan Lists

Creating and setting up scan lists on Baofeng radios is a crucial process for efficient monitoring and access to specified frequencies or channels. A scan list allows the radio to cycle through predefined channels or frequencies, enabling users to monitor multiple channels without manual selection.

**Importance of Setting Up Scan Lists:**

### 1. Enhanced Monitoring Capability

Setting up scan lists significantly enhances the monitoring capability of Baofeng radios. Instead of manually selecting and monitoring individual channels, scan lists allow users to monitor multiple channels or frequencies automatically. This feature is invaluable in situations requiring continuous monitoring of various channels, such as public safety operations or emergency response scenarios.

A well-configured scan list ensures that users can monitor critical channels without the need for constant manual switching, providing a comprehensive overview of ongoing communication activities.

### 2. Increased Situational Awareness

Scan lists play a pivotal role in enhancing situational awareness. By continuously scanning through predefined channels, users can gather information from multiple sources simultaneously. This increased situational awareness is particularly vital in dynamic environments, allowing users to stay updated on various communications and respond promptly to emerging situations.

The ability to monitor multiple channels without missing critical transmissions ensures that users have a comprehensive understanding of the operational environment.

### 3. Efficient Access to Relevant Channels

Efficient access to relevant channels is a key benefit of scan lists. By cycling through a predefined list of channels, users gain swift access to critical or frequently used frequencies. This feature is advantageous in scenarios where immediate access to specific channels, such as emergency services or primary communication channels, is crucial.

Scan lists streamline the process of accessing essential frequencies, reducing the time required to locate and select the necessary channels manually.

### 4. Minimized Missed Communications

One of the primary advantages of scan lists is the reduction in missed communications. Continuous scanning through specified channels minimizes the risk of missing important transmissions or updates. This feature ensures comprehensive coverage, reducing the likelihood of overlooking critical information or communications.

By scanning through a designated list of channels, users can effectively monitor multiple frequencies, minimizing the chances of missed communications that could impact operational effectiveness.

**Methods for Setting Up Scan Lists:**

**Creation of Scan List**

- **Identifying Channels:** Determine the channels or frequencies to include in the scan list based on their relevance and importance. Prioritize critical channels or those frequently used in specific operations or situations.
- **Grouping Channels:** Group channels logically based on their function, priority, or geographic relevance. This grouping ensures a systematic approach to scanning, allowing users to focus on specific categories of channels.
- **Programming Software:** Utilize programming software such as CHIRP to create and manage scan lists efficiently. Programming software offers an intuitive interface to organize, label, and arrange channels into scan lists on a computer, simplifying the process before transferring them to the radio.

**Configuring Scan Settings**

- **Scan Parameters:** Configure scan parameters such as the scan mode (e.g., VFO or Memory), scan delay time, and scan resume options. Adjusting these settings allows users to customize the scanning process based on their preferences and operational needs.

- **Frequency Bandwidth:** Set the frequency bandwidth to scan specific frequency ranges of interest. This customization enables users to focus on particular frequency bands relevant to their communication requirements.
- **Priority Channel:** Designate priority channels within the scan list to ensure immediate monitoring when activity is detected. Assigning priority to critical channels ensures prompt attention to important communications.

## Managing Scan Lists

- **Editing and Updating:** Regularly review and update scan lists to accommodate changes in communication needs or frequency usage. Remove obsolete channels and add new ones as necessary to ensure the scan list's relevance and effectiveness.
- **Backup and Documentation:** Consider creating backups of scan lists for future reference or in case of radio resets. Document the configurations and modifications made to the scan lists to facilitate easy restoration or modifications in the future.

Continuously managing and updating scan lists ensures their effectiveness in addressing communication requirements and adapting to evolving operational needs.

## Operational Impact of Scan Lists:

## Enhanced Operational Efficiency

- **Streamlined Monitoring:** Scan lists streamline the monitoring process by automating the scanning of specified channels. This automation reduces the need for manual intervention, allowing users to stay informed without constant interaction with the radio.
- **Quick Access to Critical Channels:** Rapid access to designated priority channels ensures immediate attention to critical communications. This feature is particularly beneficial in emergency situations or high-paced operational environments.

## Situational Awareness and Preparedness

- **Comprehensive Coverage:** Continuous scanning across multiple channels enhances situational awareness by providing a broad view of ongoing communications. It helps users stay informed about various communication activities simultaneously.
- **Preparation for Dynamic Environments:** Scan lists prepare users for dynamic situations by providing access to a range of frequencies. This readiness ensures adaptability and preparedness for changing circumstances or emerging communication needs.

**Reduced Missed Communications**

- **Minimized Missed Transmissions:** Continuous scanning through specified channels reduces the risk of missing important communications. This comprehensive coverage minimizes the likelihood of overlooking critical updates or transmissions.

**Adaptability and Customization**

- **Tailored Monitoring:** Customizable scan lists allow users to tailor monitoring based on their specific communication needs. This customization ensures a personalized and efficient scanning experience.

By leveraging scan lists effectively, Baofeng radios can significantly enhance operational efficiency, situational awareness, and communication effectiveness in various operational settings.

## Using CHIRP Software

CHIRP is a free, open-source software used for programming amateur radios. It supports a wide range of radio models, including Baofeng radios, and allows users to manage and organize radio settings, frequencies, channels, and other configurations conveniently from a computer.

**Features and Functionalities of CHIRP**

- **Radio Compatibility:** CHIRP supports a wide range of radio models, including multiple versions of Baofeng radios such as the UV-5R series, BF-F8HP, and many others. Its compatibility with various radio brands and models makes it a versatile tool for radio enthusiasts.
- **User-Friendly Interface:** The software offers an intuitive and easy-to-navigate interface, enabling users, including beginners, to program and manage their radios without extensive technical knowledge. The interface layout organizes functions logically, making it accessible for users to perform programming tasks efficiently.
- **Programming Capabilities:** CHIRP allows users to program a plethora of radio settings, including frequencies, channel names, squelch levels, transmit power levels, duplex settings, and much more. It simplifies the process of inputting and organizing these settings, making radio programming accessible to a wider audience.
- **Bulk Editing and Cloning:** One of CHIRP's significant advantages is its ability to perform bulk editing of channels and settings. Users can create, modify, or clone multiple channels simultaneously, enhancing efficiency when programming a large number of frequencies or radios with similar configurations.
- **Importing and Exporting Data:** The software facilitates the import and export of data in various formats, including CSV files. This feature allows users to manage frequencies and

channel lists outside the software and import them into CHIRP, streamlining the programming process.

- **Cross-Platform Compatibility:** CHIRP is compatible with multiple operating systems, including Windows, macOS, and Linux distributions. This cross-platform support ensures accessibility for users across different operating environments.

**Practical Applications and Benefits of Using CHIRP**

1. **Efficient Programming:** CHIRP simplifies the programming process by providing a centralized platform to input and organize radio settings. Users can swiftly program frequencies, labels, and configurations, reducing the time required for manual programming through the radio's interface.

2. **Customization and Organization:** The software allows for comprehensive customization and organization of channels. Users can assign names, categorize channels by function or priority, and arrange them systematically for ease of use. This customization enhances operational efficiency by tailoring the radio to specific communication needs.

3. **Managing Multiple Radios:** For radio enthusiasts, emergency responders, or organizations with multiple Baofeng radios, CHIRP facilitates efficient management by enabling users to program and manage multiple radios simultaneously. This bulk programming capability simplifies the configuration process for numerous radios.

4. **Community Database Integration:** CHIRP integrates a community-driven database of frequencies and channels, enabling users to access a vast repository of pre-programmed channels. Users can download and import frequencies for various regions, emergency services, or amateur radio bands, enhancing convenience and saving time on manual input.

5. **Firmware Updates:** The software supports firmware updates for compatible radios. This feature ensures that users can easily update their radios' firmware to access new functionalities or bug fixes provided by the manufacturer.

6. **Learning and Education:** CHIRP serves as an educational tool for amateur radio enthusiasts or individuals learning about radio programming. Its user-friendly interface and comprehensive functionalities provide a platform for users to explore and understand radio programming concepts.

**Steps to Use CHIRP Software for Baofeng Radios**

**Downloading and Installing CHIRP**

- **Navigate to CHIRP's Official Website:** Begin by visiting the official CHIRP website (https://chirp.danplanet.com/) to download the software. Choose the appropriate version compatible with your operating system (Windows, macOS, or Linux).
- **Installation Process:** Once downloaded, follow the installation instructions provided by CHIRP. The installation is typically straightforward, and the software's setup wizard guides users through the process.

**Connecting Baofeng Radio to the Computer**

- **Choose a Compatible Programming Cable:** Select a programming cable compatible with your Baofeng radio model. Ensure the cable's compatibility and functionality with CHIRP.
- **Connect the Radio and Computer:** Plug one end of the programming cable into the Baofeng radio's programming port and the other end into an available USB port on your computer.

**Launching CHIRP Software**

- **Open CHIRP:** Launch the CHIRP software on your computer. The software will prompt you to select the radio's manufacturer and model. Choose the appropriate Baofeng radio model from the list.

**Reading Data from the Radio**

- **Read from Radio:** Select the option to read data from the radio within the CHIRP interface. This action retrieves the existing settings, channels, and frequencies stored in the Baofeng radio's memory.
- **Retrieve Radio Configuration:** CHIRP will display the current configuration of the Baofeng radio, including programmed channels and settings. This information will be visible within the software's interface.

**Modifying or Adding Channels**

- **Modify Existing Channels:** Review and modify existing channels or frequencies as needed. Users can change channel names, frequencies, tones, offsets, and other settings within CHIRP's interface.
- **Add New Channels:** Users can also add new channels by inputting frequencies, assigning channel names, and configuring settings according to their communication requirements.

### Organizing Channels and Settings

- **Arrange Channels:** Organize the programmed channels in a logical sequence within CHIRP. Group similar channels together, label them appropriately, and arrange them in a manner that suits your operational needs.
- **Set Scan Lists:** Configure scan lists within CHIRP if desired, specifying channels to include in the scan and their priority or scanning sequence.

### Writing Data to the Radio

- **Write Data Back to the Radio:** Once modifications are made and channels organized, select the option to write data back to the Baofeng radio from within CHIRP.
- **Confirm Writing Data:** CHIRP will prompt for confirmation before writing data to the radio. Verify the settings and channels to be written and proceed with the write operation.

### Verification and Testing

- **Disconnect Programming Cable:** Safely disconnect the programming cable from the Baofeng radio once the data has been successfully written.
- **Test Radio Functionality:** Power on the Baofeng radio and navigate through the programmed channels to ensure that the frequencies, names, and settings have been correctly saved and are operational.

### Backup and Documentation

- **Create Backup Files:** Within CHIRP, consider creating backup files of the programmed settings on your computer. This backup can be useful in case the radio needs resetting or for programming other radios similarly.
- **Document Settings:** Document the programmed channels, frequencies, and specific configurations for future reference or in case of reprogramming needs.

### Best Practices and Tips for Using CHIRP:

- **Backup Your Data:** Regularly backup radio settings and configurations created or modified using CHIRP. Storing backups on your computer ensures data security and facilitates restoration in case of accidental data loss or radio resets.
- **Verify Settings Before Writing:** Before writing data to the radio, thoroughly review and verify the programmed settings within CHIRP. Ensure accuracy in frequencies, labels, and configurations to prevent errors or incorrect programming.
- **Explore Community Resources:** Utilize community-driven resources and databases available within CHIRP to access pre-programmed channels or frequencies. This resource can save time and provide valuable information for programming specific channels or bands.

- **Keep Software Updated:** Regularly check for software updates to benefit from new features, improvements, and compatibility enhancements. Keeping CHIRP updated ensures optimal performance and access to the latest functionalities.

# *Chapter 7:* **Making Your First Contact**

Whether you're a beginner or an experienced radio enthusiast, making your first contact with a Baofeng radio can be an exciting and rewarding experience. In this chapter, we will explore the steps involved in making your first contact, including finding local repeaters, calling and responding to CQ, and adhering to proper radio etiquette.

## Finding Local Repeaters

Finding local repeaters is an essential step in amateur radio operation, particularly when using a Baofeng radio. These repeaters serve as crucial communication hubs that amplify signal range, enabling long-distance communication within a designated area.

### Importance of Locating Local Repeaters

#### 1. Extended Communication Range

One of the primary advantages of utilizing local repeaters is the extension of communication range. Unlike direct transmissions, which are limited by line-of-sight constraints, repeaters receive signals and retransmit them at higher power levels, typically from elevated locations. This process allows signals to overcome obstacles such as buildings, hills, and other obstructions that would otherwise hinder communication. By leveraging local repeaters, operators can effectively communicate over distances that would be unattainable through direct transmissions alone. This expanded range opens up opportunities for connecting with other operators in neighboring towns, regions, or even countries.

#### 2. Improved Signal Quality

In addition to extending the communication range, local repeaters often enhance the overall signal quality. As signals are received and retransmitted by the repeater, they undergo amplification, filtering, and noise reduction processes. This helps to mitigate signal degradation, interference, and background noise that may be present in the original transmission. Consequently, operators benefit from clearer and more reliable communications, ensuring that their messages are delivered with greater clarity and intelligibility.

#### 3. Increased Reliability and Redundancy

Local repeaters contribute to the overall reliability and redundancy of amateur radio operations. In emergency situations or during adverse environmental conditions, such as severe weather events, repeaters can serve as vital communication links. They can be strategically positioned to provide coverage over large areas or specific regions, enabling operators to stay connected and informed even when other forms of communication may be compromised. The redundancy offered by multiple repeaters within an area ensures that if one repeater experiences technical issues or becomes unavailable, operators can quickly switch to an alternative repeater for continued communication.

### 4. Access to Local Amateur Radio Networks

Many local repeaters are linked to amateur radio networks, which facilitate broader communication capabilities. These networks connect repeaters over larger geographic areas, enabling operators to communicate with individuals across a network of repeaters. By accessing local repeaters, operators gain entry into these networks, allowing them to participate in scheduled nets, emergency communications exercises, and other collaborative activities. These networks foster a sense of community among operators and provide opportunities for learning, information exchange, and support within the amateur radio community.

### 5. Emergency Communication

Local repeaters play a vital role in emergency communication scenarios. During natural disasters, public emergencies, or other critical situations, amateur radio operators often step in to provide essential communication services when conventional systems fail or become overloaded. Local repeaters, especially those linked to emergency communication networks, serve as vital communication hubs in such situations. They enable operators to relay crucial information, coordinate response efforts, and provide assistance to affected individuals or communities. The ability to locate and utilize local repeaters is essential for operators who wish to contribute effectively during emergencies and provide valuable support to their communities.

### 6. Learning and Skill Development

Locating and utilizing local repeaters also offers valuable learning opportunities for amateur radio operators. By actively participating in repeater networks, operators can engage with experienced hams, learn from their expertise, and gain insights into best practices, operating techniques, and equipment recommendations. Joining nets or monitoring repeater traffic exposes operators to a wide range of communication styles, protocols, and emergency communication procedures. The interaction within the amateur radio community fosters personal growth, skill development, and the exchange of knowledge among operators.

**Methods to Find Local Repeaters**

Here, we will discuss various methods and resources to help you locate repeaters in your area.

### 1. Online Databases

One of the most convenient ways to find local repeaters is by using online databases. These databases provide comprehensive lists of repeaters, including their frequencies, locations, and other relevant information. Here are a few popular online databases:

- **Repeaterbook.com:** Repeaterbook is a widely used online database that allows users to search for repeaters worldwide. You can search by location, frequency range, and other parameters. The database provides detailed information about each repeater, including its coordinates, callsign, offset direction, and tone (if required).
- **Radio Amateurs of Canada (RAC) Website:** If you are in Canada, the RAC website is an excellent resource for finding local repeaters. They maintain an up-to-date repeater directory that includes information on repeaters across the country.
- **National Amateur Radio Society Websites**: Many countries have national amateur radio societies or organizations that maintain repeater directories on their websites. These directories often provide accurate and reliable information about local repeaters. Examples include the American Radio Relay League (ARRL) in the United States and the Radio Society of Great Britain (RSGB) in the UK.

When using online databases, it's important to verify the information and cross-reference it with multiple sources. Some databases rely on user submissions, which may not always be up to date. Therefore, it's a good practice to double-check the repeater details before programming them into your Baofeng radio.

### 2. Amateur Radio Clubs and Organizations

Another valuable resource for finding local repeaters is amateur radio clubs and organizations. These groups often maintain and operate repeaters in their respective areas. Reach out to local clubs or organizations and inquire about repeaters in your vicinity. They can provide you with accurate and current information, including repeater frequencies, access requirements, and any specific guidelines for using their repeaters.

Joining a local amateur radio club not only helps you in finding repeaters but also provides opportunities to connect with experienced operators who can share their knowledge and assist you in various aspects of amateur radio.

### 3. Mobile Applications

Smartphone applications can be convenient tools for locating local repeaters on the go. Some popular apps include:

- **RepeaterBook:** The RepeaterBook app, available for both Android and iOS devices, provides access to the Repeaterbook.com database. It uses your device's GPS to locate nearby repeaters and displays their details, making it easy to find repeaters while traveling.
- **ARRL Repeater Directory:** The ARRL Repeater Directory app, developed by the American Radio Relay League, provides access to the ARRL repeater database. It offers search functionality based on location or frequency and includes information about repeaters in the United States and Canada.

These apps are user-friendly and allow you to search for repeaters based on location, frequency, or other criteria. They often provide additional features such as mapping, filtering, and the ability to save favorite repeaters for quick reference.

### 4. Local Ham Radio Events and Nets

Attending local ham radio events, such as field days or hamfests, can provide opportunities to learn about local repeaters and connect with experienced operators in your area. These events often have information booths or knowledgeable individuals who can guide you in finding and utilizing local repeaters effectively.

Participating in organized nets (scheduled on-air meetings) is another way to gather information about local repeaters. Nets often include discussions about available repeaters and their coverage areas. Net participants can provide valuable insights and recommendations based on their experiences.

### Programming Local Repeaters into Baofeng Radios

Once you have identified local repeaters in your area, the next step is to program them into your Baofeng radio. Programming repeater frequencies allows you to easily access and communicate through these repeaters, expanding your range and connectivity in the amateur radio world. In this section, we will guide you through the process of programming local repeaters into your Baofeng radio.

### 5. Obtain Repeater Information

Before programming repeaters into your Baofeng radio, you will need the following information:

- **Repeater Frequency:** The frequency at which the repeater operates. This is usually provided in megahertz (MHz).
- **Offset Direction:** The offset direction determines the difference between the repeater's transmit and receive frequencies. Common offset directions include positive (+), negative (-), or no offset (simplex). For example, if the repeater has a positive offset of 600 kHz, the transmit frequency will be higher than the receive frequency by 600 kHz.
- **Offset Frequency:** The amount by which the transmit frequency differs from the receive frequency. This value is usually given in kilohertz (kHz). For example, if the receive frequency is 147.000 MHz and the offset frequency is 600 kHz, the transmit frequency will be 147.600 MHz.
- **Tone (CTCSS/DCS):** Some repeaters require a specific tone for access, known as Continuous Tone-Coded Squelch System (CTCSS) or Digital-Coded Squelch (DCS). If the repeater requires a tone, you will need to determine the appropriate CTCSS or DCS code.

6. **Accessing Programming Mode**

To program repeaters into your Baofeng radio, you need to access the programming mode. Here's a general overview of the steps:

- Turn on your Baofeng radio.
- Press the MENU button to enter the menu mode.
- Use the arrow buttons or channel selector knob to navigate to the programming menu option. It is usually labeled as "Program," "VFO/MR," or similar.
- Press the MENU button again to enter the programming mode.

7. **Programming Repeater Frequencies**

Once you are in the programming mode, follow these steps to program repeater frequencies into your Baofeng radio:

- **Select an empty memory channel:** Use the arrow buttons or channel selector knob to navigate to an empty memory channel where you want to store the repeater frequency.
- **Enter the repeater frequency:** Use the numeric keypad to enter the repeater's receive frequency. Be sure to include the decimal point.
- **Set the offset direction:** Use the arrow buttons or channel selector knob to navigate to the offset direction option. Select the appropriate offset direction based on the repeater's specifications: positive (+), negative (-), or simplex (no offset).
- **Set the offset frequency:** Navigate to the offset frequency option and enter the offset frequency in kilohertz (kHz).

- **Enable or disable the CTCSS/DCS tone:** If the repeater requires a CTCSS or DCS tone, navigate to the CTCSS/DCS option and select the appropriate tone code. If no tone is required, select "Off" or "None."
- **Save the programmed channel:** Once you have entered all the necessary information, save the programmed channel by pressing the MENU button or a dedicated save button, if available.

## 8. Testing and Using Programmed Repeater Channels

After programming the repeater frequencies into your Baofeng radio, it's essential to test and ensure they work correctly. Here are some additional steps to follow:

- **Select the programmed repeater channel:** Use the arrow buttons or channel selector knob to navigate to the channel where you programmed the repeater frequency.
- **Transmit and listen:** Press the PTT (Push-To-Talk) button and transmit a test signal while listening for any responses or activity on the repeater's output frequency.
- **Adjust squelch and volume:** If you receive a signal, adjust the squelch level to eliminate background noise and adjust the volume to a comfortable level.
- **Follow proper radio etiquette:** When communicating through the repeater, remember to adhere to proper radio etiquette, including identifying yourself, waiting for your turn to speak, and maintaining a professional tone.

**Benefits of Leveraging Local Repeaters:**

- **Enhanced Communication Capabilities**: Utilizing local repeaters significantly enhances the communication capabilities of Baofeng radios. It extends the range, improves signal quality, and provides more reliable communications over longer distances.
- **Community Engagement and Networking:** Accessing local repeaters opens doors to engaging with a diverse community of amateur radio enthusiasts. It facilitates interactions, knowledge sharing, and participation in various activities within the amateur radio community.
- **Emergency Communication Preparedness:** Local repeaters often play a vital role in emergency communication scenarios. They serve as key hubs for coordinating emergency responses, disseminating critical information, and providing communication channels during crises or natural disasters.

## Calling and Responding to CQ

CQ is a standard calling signal used in amateur radio to initiate communication with other operators. It is derived from the French word "sécurité" and is universally recognized as an invitation to engage in a conversation. When an operator wants to establish contact with other operators, they will transmit the call "CQ" followed by additional information like their call sign, location, or any other relevant details.

To respond to a CQ call, an operator listens for the CQ transmission and, if interested in initiating contact, will transmit their own call sign followed by the call sign of the station they are responding to. This exchange allows for the establishment of a two-way communication link between the operators.

Calling and responding to CQ is a common practice in amateur radio, especially during contests or when operators are seeking contacts with other stations. It serves as a way to connect with fellow operators, exchange information, and engage in conversations about various radio-related topics.

It's important to note that when calling or responding to CQ, operators should adhere to proper radio etiquette, follow frequency and power regulations, and maintain a professional and respectful tone during the communication.

## Calling CQ

### 1. Understanding the Purpose of CQ

The primary purpose of calling CQ is to initiate a contact with other amateur radio operators. It is an invitation for someone to respond and engage in a conversation. When calling CQ, you are essentially announcing your presence and availability for communication. It is an excellent way to connect with operators in your local area or even around the world.

### 2. Selecting the Frequency and Band

Before calling CQ, it is essential to select an appropriate frequency and band based on your equipment capabilities and the desired reach of your communication. Choose a frequency that aligns with the band conditions and is within the allowed amateur radio bands for your license class. Avoid interfering with ongoing conversations or established nets on that frequency.

### 3. Listening Before Calling

Before initiating a CQ call, it is good practice to listen to the frequency for a while to ensure it is clear and not already in use by other operators. Listening allows you to be aware of ongoing conversations and prevents unintentional interruptions. It also helps you gauge the activity level on the frequency and determine if it is suitable for your call.

4. **Structuring Your CQ Call**

When calling CQ, it is important to structure your call in a clear and concise manner. Here is a typical format for a CQ call:

- **Start with "CQ" repeated three times:** This repetition helps catch the attention of operators who may be listening casually or tuning across the band.
- **Follow "CQ" with your call sign:** State your call sign clearly, phonetically if necessary, and repeat it two to three times. This allows other operators to identify and respond to your call.
- **Mention the frequency and band:** After your call sign, state the frequency and band you are operating on. For example, "CQ, CQ, CQ, this is [your call sign] calling on [frequency] in the [band]."
- **Optional:** Include additional information: Depending on your preference and the situation, you can add additional information to your CQ call, such as your location, power output, or any specific topics you are interested in discussing. Keep this information brief and relevant.

5. **Repeat Your CQ Call**

After structuring your initial CQ call, it is important to repeat it at regular intervals to increase the chances of being heard by operators who may have missed the initial call. Space out your repetitions adequately, allowing time for potential responders to acknowledge and join the conversation.

6. **Listening for Responses**

While repeating your CQ call, actively listen for responses from other operators. They may call your call sign or indicate their interest in making contact with you. Be attentive and patient, as it may take some time for responses to come in, especially if the frequency is not heavily populated.

7. **Responding to Calls**

When you receive a response to your CQ call, promptly acknowledge the calling station by using their call sign. For example, if someone responds with "This is [their call sign]," reply with "Hello [their call sign], this is [your call sign]." This establishes a connection and confirms the successful reception of their call.

## Responding to CQ

### 8. Active Listening

When actively listening for CQ calls, it is important to pay close attention to the frequency you are monitoring. Be mindful of ongoing conversations and nets, and avoid interfering with them. Focus on identifying CQ calls by their distinct repetitive pattern, usually starting with "CQ" followed by a call sign.

### 9. Confirming the CQ Call

Once you hear a CQ call, verify that the call sign is directed toward your station by listening for your own call sign. The operator initiating the CQ call may be specifically addressing your station or calling out to a broader audience. If the call sign matches yours, it indicates an opportunity for you to respond.

### 10. Structuring Your Response

When responding to a CQ call, structure your response in a clear and concise manner. Here is a typical format for responding to a CQ call:

- **Start with the calling station's call sign:** Begin your response by stating the call sign of the station that initiated the CQ call. For example, "This is [their call sign], [their call sign], [their call sign], this is [your call sign]."
- **Follow with your signal report:** Provide a brief signal report to indicate the quality of the received signal. This report typically includes a numerical value between 1 and 5, with 5 being the strongest. For example, "You are 5 by 9, 5 by 9, over."
- **Introduce yourself:** After the signal report, state your own call sign, phonetically if necessary, and repeat it two to three times. This allows the calling station to identify and acknowledge your response. For example, "My call sign is [your call sign], [your call sign], [your call sign]."
- **Optional:** Add additional information: Depending on the situation and your preferences, you can include additional information in your response, such as your location, equipment setup, or any specific topics you would like to discuss. Keep this information brief and relevant.

### 11. Wait for Acknowledgment

After sending your response, give the calling station time to acknowledge and respond to your call. Be patient and avoid transmitting while the other station is speaking. Listen attentively for their acknowledgment and any further instructions or questions they may have.

### 12. Engaging in the Conversation

Once the calling station acknowledges your response and indicates their interest in continuing the conversation, you can engage in a meaningful exchange. Follow standard communication practices, such as taking turns speaking, allowing pauses for responses, and being courteous and respectful in your interactions.

### 13. Ending the Contact

When you and the other operator have concluded your conversation, it is time to end the contact. Express your farewell and gratitude for the communication. You can use phrases like "Thank you for the contact, [their call sign]. It was a pleasure talking with you. 73 and best wishes."

## Proper Radio Etiquette

Proper radio etiquette is crucial for effective and respectful communication. By following established guidelines, operators can ensure clear transmissions, minimize interruptions, and maintain a professional atmosphere.

### Importance of Radio Etiquette

#### 1. Clarity and Precision

Effective communication hinges on clarity. Proper radio etiquette ensures messages are transmitted clearly and concisely, reducing the chances of misinterpretation or confusion. In scenarios like emergency response or critical operations, clear communication is paramount for successful outcomes. It enables accurate conveyance of vital information, minimizing errors and enhancing coordination.

Proper radio etiquette plays a vital role in ensuring clarity and precision during communication. By following established guidelines, users can effectively transmit their messages in a clear and concise manner. This is particularly important in situations where miscommunication or confusion can have severe consequences, such as emergency response operations or critical industrial processes. Clear communication allows for the accurate conveyance of information, ensuring that instructions, updates, and requests are understood correctly by all parties involved. By reducing the chances of errors or misunderstandings, proper radio etiquette enhances coordination and improves overall operational outcomes.

## 2. Respect for Others

Etiquette fosters an environment of respect among radio users. It emphasizes fair use of communication channels, preventing unnecessary interruptions and ensuring equitable opportunities for all to transmit their messages without disruptions. Respecting others' airtime ensures that each user has a chance to convey their information efficiently, contributing to smoother and more effective communication.

Respect for others is a fundamental aspect of proper radio etiquette. By adhering to established guidelines, users demonstrate consideration for their fellow communicators. This includes avoiding unnecessary interruptions or disruptions that could hinder the flow of communication. By allowing each user to transmit their messages without undue interference, proper radio etiquette ensures that all participants have an equal opportunity to be heard. This not only promotes fairness but also contributes to more efficient and effective communication overall.

## 3. Efficiency and Effectiveness

Following established etiquette guidelines streamlines communication processes. Clear, succinct messages enable quicker decision-making, especially in time-sensitive situations where prompt action is necessary, such as coordinating rescue efforts or managing security operations. Efficient communication enhances productivity and ensures that information is conveyed swiftly and accurately to relevant parties.

Efficiency and effectiveness are key benefits of proper radio etiquette. By adhering to established guidelines, users can streamline their communication processes, leading to more efficient exchanges. Clear and concise messages allow for quicker decision-making, particularly in time-sensitive situations where rapid responses are crucial. By eliminating unnecessary or extraneous information, proper radio etiquette ensures that messages are transmitted swiftly and accurately, allowing for timely actions and responses. This enhances overall productivity and contributes to the success of various operations and activities that rely on radio communication.

## 4. Professionalism

In professional settings, adherence to proper etiquette demonstrates professionalism. Whether in emergency services, security operations, or amateur radio use, maintaining a respectful and disciplined approach to communication enhances credibility and trust among peers. It instills confidence in the ability to handle situations competently and reliably.

Professionalism is a key aspect of proper radio etiquette, particularly in professional settings. By adhering to established guidelines, users demonstrate their commitment to professionalism and the importance they place on effective communication. This is particularly relevant in fields such as emergency services and security operations, where clear and reliable communication is vital for ensuring safety and successful outcomes. By following proper radio etiquette, individuals convey a sense of competence and professionalism, instilling confidence in their peers and contributing to a positive work environment.

5. **Error Reduction and Clarity**

Proper etiquette significantly decreases the likelihood of errors or confusion in communication. This is vital, particularly when relaying critical information or instructions that could impact safety or operational success. Clear and concise messages reduce the chances of misunderstandings, ensuring that information is conveyed accurately and effectively.

Proper radio etiquette plays a crucial role in reducing errors and improving clarity in communication. By adhering to established guidelines, users can minimize the chances of misunderstandings or misinterpretations, particularly when conveying critical information or instructions. Clear and concise messages, free from unnecessary distractions or ambiguities, ensure that information is accurately transmitted and understood. This is particularly important in situations where the consequences of errors or misunderstandings can be significant, such as emergency response operations or complex industrial processes. By following proper radio etiquette, users contribute to the overall clarity and effectiveness of communication, reducing the likelihood of errors and improving operational outcomes.

**Basic Rules of Radio Etiquette**

6. **Identify Yourself**

Each transmission should begin with a clear identification or call sign. This practice ensures transparency in communication and enables other users to recognize the sender immediately, preventing confusion or ambiguity. Identifying oneself is not just a formality but a crucial aspect of effective communication, especially in group discussions or emergency situations.

Identifying oneself at the beginning of each transmission is a fundamental rule of radio etiquette. This allows other users to immediately recognize the sender and establishes transparency in communication. By clearly stating one's identification or call sign, confusion and ambiguity can be avoided, particularly in situations involving group discussions or emergency response operations. Proper identification ensures that messages are attributed to the correct sender, facilitating efficient and effective communication.

### 7. Listen Before Transmitting (LBT)

Engaging in "listening before talking" is fundamental in proper radio etiquette. This practice involves taking the time to listen to ongoing conversations before transmitting any messages. By listening before speaking, users can avoid accidental interference with ongoing discussions and gain a better understanding of the current context. This allows for more organized and coherent exchanges, as users can determine the appropriate timing and relevance of their communication. Active listening also helps in understanding the flow of conversation and enables users to respond appropriately and effectively.

### 8. Speak Clearly and Slowly

Enunciation and a moderate speaking pace are pivotal in proper radio etiquette. By speaking clearly and slowly, users ensure that their messages are comprehensible even in challenging or noisy environments. This is particularly important when transmitting critical information or instructions that need to be accurately understood. Clear and deliberate speech reduces the chances of misunderstanding or misinterpretation, allowing for effective communication even in less-than-ideal conditions.

### 9. Avoid Jargon and Codes

While technical terms might be necessary in certain contexts, excessive use of jargon can hinder understanding in radio communication. Proper radio etiquette emphasizes the use of clear and simple language to ensure that messages are universally understood. When technical terms are used, it's essential to ensure that they are widely recognized among the users in the communication network. This promotes clarity and facilitates effective communication among all parties involved.

### 10. Keep Messages Brief and Relevant

Transmitting only essential information and avoiding unnecessary details is an important rule of radio etiquette. By keeping messages brief and relevant, users enhance the efficiency of communication. Conciseness allows for quicker responses and actions, as recipients can quickly process the information without being overwhelmed by irrelevant or extraneous details. Transmitting only pertinent information ensures that the message is received and understood without confusion or distractions, contributing to effective and streamlined communication.

## 11. Wait for Confirmation

After sending a message, patience is crucial in proper radio etiquette. Users should wait for acknowledgment or confirmation before assuming that their message was received and understood. This practice ensures that important information is not overlooked or misunderstood, preventing potential miscommunication. Waiting for confirmation also allows for the acknowledgment of the received message, ensuring that the communication loop is complete and that all parties are on the same page.

## 12. Respect Channel Usage

Proper channel management is essential in radio communication. Users should avoid unnecessary channel hopping or monopolizing specific channels, ensuring fair access for everyone and minimizing disruptions. Respecting channel usage promotes a more organized and efficient communication environment. It allows for smoother exchanges and prevents unnecessary interference or congestion on the channels. By adhering to proper channel usage, users contribute to the overall effectiveness and professionalism of radio communication.

## 13. Emergency Protocol

Understanding and following emergency procedures is critical in proper radio etiquette. During emergencies, prioritizing emergency transmissions and conveying information clearly and swiftly are paramount for effective response and resolution. Being well-versed in emergency protocols ensures a rapid and coordinated response in critical situations, potentially saving lives and resources. Properly following emergency procedures allows for the efficient and effective utilization of radio communication during emergency situations, facilitating timely and accurate information exchange among responders.

By following these basic rules of radio etiquette, users can ensure clear, efficient, and effective communication in various contexts. Proper radio etiquette promotes professionalism, minimizes errors and confusion, and enhances overall communication outcomes. Whether in emergency response, security operations, or everyday radio use, adhering to these etiquette guidelines fosters a culture of respect and proficiency in radio communication.

# *Chapter 8:* **Troubleshooting and Maintenance**

Baofeng radios are reliable communication devices, but like any electronic equipment, they may encounter issues from time to time. Understanding how to troubleshoot common problems, properly maintain your device, and manage firmware updates are essential for ensuring optimal performance.

## Common Issues and Solutions

### 1. Poor Signal Reception or Transmission

**Issue:** Weak signal reception or transmission can significantly impact effective communication.

**Solution:**

- **Antenna Inspection and Upgrades:** Start by inspecting the antenna. Ensure it is properly attached and undamaged. A compromised or improperly connected antenna can drastically reduce signal strength. Consider upgrading to a higher-gain or better-quality antenna for improved performance, especially in areas with weak signals.
- **Signal Optimization:** Experiment with the radio's positioning and orientation. Moving to a higher location or adjusting the radio's angle might enhance signal reception. Additionally, being aware of the line of sight and potential obstructions between transmission points can aid in optimizing signal strength.
- **External Antenna Usage:** For increased range and improved signal quality, consider using an external antenna. External antennas, especially those mounted at greater heights or outdoors, can significantly enhance the radio's performance, especially in long-range communication scenarios.
- **Squelch Adjustment and Filtering:** Adjust the squelch settings appropriately to filter out unwanted noise without blocking weaker signals. Understanding how squelch works and adjusting it to the right level is crucial for optimizing signal reception.
- **Signal Booster Utilization:** In situations with consistently weak signals, using signal boosters or amplifiers can be beneficial. Signal boosters increase the strength of received signals, thereby improving overall reception quality.

### 2. Audio Problems

**Issue:** Distorted sound, low volume, or complete absence of audio can hinder effective communication.

**Solution:**

- **Volume Management:** Ensure the volume is set to an audible level. Sometimes, users inadvertently adjust the volume to very low levels, leading to audio issues. Properly adjusting the volume knob or settings can often resolve this problem.
- **Examine Speaker and Earpiece:** Inspect the speaker or earpiece for any blockages or damage. Dust, debris, or physical damage to these components can result in distorted or no audio output. Cleaning or replacing these components as needed can rectify the issue.
- **Accessories Testing:** Test different audio accessories such as earpieces, external speakers, or headphones. This process helps isolate whether the problem lies with the accessory or the radio itself. Using known working accessories aids in diagnosing the issue accurately.
- **Channel Verification and Audio Settings:** Ensure the selected channel has activity and is within the range of transmission. Additionally, verifying the audio settings of the radio, such as volume, tone, or squelch settings, can help identify and rectify audio-related issues.
- **Environmental Considerations:** Be mindful of environmental factors affecting audio quality. Interference from nearby electronic devices, atmospheric conditions, or physical barriers can impact audio transmission and reception. Minimizing interference sources and adjusting settings accordingly can mitigate these issues.

3. **Programming Errors**

**Issue:** Incorrectly programmed channels or failed programming attempts can impede effective communication.

**Solution:**

- **Thorough Review of Settings:** Double-check all entered frequencies, offsets, tones, and settings during programming. Even a minor mistake can lead to programming errors and communication issues.
- **Programming Software Use:** Employ user-friendly programming software like CHIRP or Baofeng's official programming software. These tools offer a graphical interface, making programming easier and less prone to errors. Moreover, they ensure compatibility between the radio model and software used for programming.
- **Compatibility Assurance:** Ensure compatibility between the radio model and the programming software or drivers being used. Mismatched software or drivers can lead to programming errors or failed attempts.
- **Factory Reset and Reprogramming:** If persistent issues arise, consider performing a factory reset on the radio. This resets the device to its default settings and resolves many

programming-related issues. Subsequently, reprogram the radio from scratch to ensure accurate settings.

- **Backup Configuration:** Maintain a backup of the radio's programming configuration. Saving a backup file ensures that if any programming errors occur, you can easily restore the device to its previous working state without starting from scratch.

## 4. Battery Drainage

**Issue:** Rapid battery depletion can disrupt communication, particularly when a reliable power source is essential.

**Solution:**

- **Quality Batteries Usage:** Always utilize genuine Baofeng batteries or certified replacements. Inferior-quality batteries may have a shorter lifespan and can potentially damage the radio. Investing in high-quality batteries ensures longer usage time and reliable performance.
- **Power Management Features Activation:** Activate power-saving or energy-efficient features on the radio. For instance, adjusting screen brightness, turning off unnecessary features, or activating power-saving modes can extend battery life significantly.
- **Regular Charging Practices:** Adhere to proper charging practices. Avoid overcharging the batteries, as it can diminish their overall capacity. Charge the batteries at room temperature and remove them from the charger once fully charged to prevent damage.
- **Backup Batteries Provision:** Always carry spare batteries or a backup power source, especially during extended usage periods or critical communication situations. Having readily available charged batteries ensures uninterrupted communication by swiftly replacing depleted ones.
- **Contact Maintenance:** Routinely inspect and clean the battery contacts. Over time, dirt, dust, or corrosion can accumulate on the contacts, hindering proper electrical connections and contributing to battery drainage. Use a soft, dry cloth or an approved contact cleaner for cleaning.
- **Battery Storage:** Properly store spare batteries in a cool, dry place away from direct sunlight. Keeping spare batteries charged and rotating their usage periodically prevents deterioration and ensures they are ready for use when needed.

## 5. Interference and Cross-Talk

**Issue:** Interference or cross-talk from nearby radios or electronic devices can disrupt communication and cause signal distortion.

**Solution:**

- **Frequency Adjustment:** If experiencing interference from nearby radios operating on similar frequencies, adjust your radio's frequency slightly to find a less congested channel. Using frequency offset or shift settings can help mitigate interference.
- **Antenna Positioning:** Experiment with the antenna's positioning. Sometimes, a simple change in the antenna's orientation or location can reduce interference.
- **Use CTCSS/DCS Tones:** Utilize Continuous Tone-Coded Squelch System (CTCSS) or Digital-Coded Squelch (DCS) tones to filter out unwanted signals. These tones restrict reception to only those signals using the same tone, reducing interference from other sources.
- **Shielding and Grounding:** Install shielding or use grounded cables for the radio's power supply or antenna connections to minimize electromagnetic interference from nearby electronic devices.

### 6. Display and Button Malfunctions

**Issue:** Problems with the radio's display or buttons, such as unresponsiveness, erratic behavior, or malfunctioning, can hinder proper operation.

**Solution:**

- **Reset and Firmware Update:** Perform a factory reset to restore default settings. Additionally, ensure the radio's firmware is up-to-date, as outdated firmware can cause operational glitches.
- **Button Inspection and Cleaning:** Inspect the buttons for any physical damage, dirt, or moisture ingress. Clean the buttons and surrounding areas using a mild cleaning solution or a damp cloth to remove debris that might be hindering their functionality.
- **Contact Maintenance:** Check and clean the contacts beneath the buttons. Over time, dirt accumulation may affect proper contact, causing button malfunctions.
- **Professional Repair:** If issues persist, seek assistance from authorized service centers or technicians experienced in repairing Baofeng radios. They can diagnose and resolve more complex hardware-related problems.

### 7. Transmitting Range Issues

**Issue:** Limited transmitting range or difficulties in reaching desired communication distances can be frustrating.

**Solution:**

- **Power Settings Check:** Ensure the radio is set to the appropriate power output. Lower power settings conserve battery but might reduce transmission range. Adjusting to higher power settings might enhance transmission distance.
- **Antenna Upgrades:** Consider using a more efficient or longer antenna to improve transmission range. A properly tuned and quality antenna can significantly enhance the radio's reach.
- **Obstruction Identification:** Identify and minimize obstacles between communication points. Buildings, terrain, or geographical features can obstruct signals. Elevating the radio or choosing a higher location can often improve transmitting range.
- **Repeaters Usage:** Utilize repeaters if available in your area. Repeaters amplify and rebroadcast signals, extending the radio's range beyond its natural limitations.

### 8. Radio Gets Hot During Use

**Issue:** The radio becomes excessively hot during prolonged use, which can be concerning and might affect its performance.

**Solution:**

- **Ventilation:** Ensure that the radio has proper ventilation. Avoid covering or obstructing the vents as overheating can occur when the device doesn't have sufficient airflow. Use the radio in well-ventilated areas.
- **Reduce Transmit Time:** Limit continuous transmitting periods. Prolonged transmission can cause the radio to heat up quickly. Allow intervals between transmissions to let the device cool down.
- **Check for Malfunctioning Components:** Inspect the internal components for any malfunctioning parts. Faulty components might cause the radio to generate excess heat. Seek professional assistance if the problem persists.

### 9. Inability to Access Specific Features or Functions

**Issue:** Users may encounter difficulties in accessing specific features or functions within the radio's menu system.

**Solution:**

- **Menu Navigation:** Review the user manual to understand the menu system thoroughly. Baofeng radios often have multiple layers of menus. Familiarizing yourself with the navigation can help access all available features.

- **Factory Reset:** Consider performing a factory reset if the issue persists. A factory reset restores the radio to its default settings, which might resolve any software-related issues that prevent access to certain features.
- **Update Firmware:** Ensure that the radio's firmware is updated to the latest version available. Sometimes, firmware updates add new functionalities or address software bugs that might hinder access to specific features.
- **Professional Assistance:** If unable to access critical functions or features, seek guidance from experienced users or Baofeng's customer support for troubleshooting steps or potential workarounds.

## 10. Microphone or PTT Button Malfunctions

**Issue:** The microphone or Push-to-Talk (PTT) button may malfunction, resulting in difficulties in transmitting or initiating communication.

**Solution:**

- **Physical Examination:** Inspect the microphone and PTT button for any physical damage, wear, or loose connections. Repair or replace these components if necessary.
- **Clean Connections:** Clean the microphone and PTT button connections. Dust or dirt accumulation can hinder proper contact, causing malfunctions. Use a soft, dry cloth or a cotton swab to clean these components.
- **Testing and Verification:** Test the microphone and PTT button by connecting an alternate microphone or speaker-microphone accessory. This helps determine if the issue lies with the microphone itself or its connection to the radio.
- **Technical Support:** If the issue persists, seek technical assistance from authorized service centers or experienced technicians. They can diagnose and resolve complex hardware-related problems with the microphone or PTT button.

## Battery Care and Replacement

Baofeng radios are versatile communication devices powered by rechargeable batteries. Proper care and maintenance of these batteries are essential to ensure optimal performance and longevity of the radio.

### Charging Practices and Guidelines

Proper charging practices play a pivotal role in maintaining the health and performance of Baofeng radio batteries. Understanding the best charging practices helps prevent battery damage and ensures longevity.

- **Use Approved Chargers:** Baofeng radios typically come with dedicated chargers. It's crucial to use these approved chargers or certified alternatives compatible with the specific battery model. Avoid using third-party chargers that may not meet the necessary specifications, potentially causing damage to the battery or radio.
- **Avoid Overcharging:** Overcharging can reduce the battery's lifespan. Once the battery is fully charged, promptly disconnect it from the charger. Continuous charging after reaching full capacity can lead to overheating and damage.
- **Charging Environment:** Charge the battery in a stable and appropriate environment. Ensure the charging area is well-ventilated and at a moderate temperature. Extreme temperatures can adversely affect the battery's performance.
- **Proper Charging Time:** Allow sufficient time for the battery to charge fully. Interruptions during the charging process or disconnecting the battery prematurely might affect its overall capacity.

**Storage Guidelines for Baofeng Radio Batteries**

Storing batteries correctly when not in use is crucial to maintain their health and prevent degradation. Following proper storage guidelines ensures the batteries remain viable and functional for extended periods.

- **Optimal Storage Conditions:** Store Baofeng radio batteries in a cool, dry place away from direct sunlight and moisture. Extreme temperatures can impact the battery's chemistry, leading to reduced performance or capacity loss.
- **Charge Level for Storage:** Before long-term storage, partially charge the battery to around 50%. Storing a battery at full capacity or completely discharged state for extended periods can deteriorate its overall health. Regularly cycling the battery by partial charging and discharging can help maintain its condition.
- **Periodic Check-ups:** Regularly inspect stored batteries. Ensure they are clean, free from corrosion, and have maintained their charge level. Charge or partially discharge and recharge stored batteries periodically to prevent capacity loss due to inactivity.

**Maintenance Tips for Baofeng Radio Batteries** Proper maintenance prolongs the lifespan and ensures the optimal performance of Baofeng radio batteries. Implementing routine maintenance practices helps keep the batteries in excellent condition.

- **Clean Battery Contacts:** Routinely inspect and clean the battery contacts using a dry, soft cloth. Dirty or corroded contacts hinder proper electrical connections, impacting the battery's performance.

- **Avoid Deep Discharges:** Limit deep discharges of the battery whenever possible. Deep discharges can stress the battery and reduce its overall lifespan. Recharge the battery before it gets too low to maintain its health.
- **Use Genuine Batteries:** Always use genuine Baofeng batteries or certified replacements. Low-quality or counterfeit batteries can potentially damage the radio and compromise safety.
- **Avoid Extreme Conditions:** Prevent exposing the batteries to extreme temperatures, whether hot or cold. Extreme temperatures can negatively impact the battery's chemistry and reduce its efficiency.

**Baofeng Radio Battery Replacement Process**

Eventually, Baofeng radio batteries will reach the end of their lifespan and require replacement. Understanding the replacement process is essential for maintaining uninterrupted communication.

- **Identify Compatible Batteries:** Ensure the replacement battery is compatible with the specific Baofeng radio model. Using incompatible batteries may damage the radio or compromise its performance.
- **Turn off the Radio:** Before replacing the battery, turn off the radio and disconnect it from any power sources. This ensures safety and prevents damage during the replacement process.
- **Carefully Remove Old Battery:** Gently remove the old battery from the radio. Avoid using excessive force to prevent damage to the battery compartment or other components.
- **Install New Battery:** Align the new battery correctly with the contacts and insert it into the battery compartment. Ensure it fits snugly and securely to establish proper electrical connections.
- **Test and Verify:** After replacing the battery, turn on the radio and test its functionality. Ensure the new battery powers the radio correctly and maintains a charge without any issues.

Note: Always prioritize safety and follow manufacturer-recommended practices outlined in the user manual for efficient battery care and replacement.

# Firmware Updates and Resets

*Firmware Updates*

Firmware updates are integral for Baofeng radios as they offer improvements, bug fixes, and enhancements.

1. **Check for Updates:** Regularly checking for updates on the Baofeng website or authorized distributor sites is essential. Staying vigilant ensures that you're aware of available firmware updates specifically designed for your Baofeng radio model. This proactive approach prevents missing critical updates that could enhance your radio's performance and features.

2. **Download the Firmware:** Identifying and downloading the correct firmware version for your Baofeng radio is paramount. Selecting the precise update that matches your device model is crucial to prevent potential compatibility issues or, worse, the risk of damaging your radio. Always acquire firmware updates from official sources to mitigate the risk of malware or corrupted files.

3. **Backup Settings and Data:** Prior to initiating the update process, creating a comprehensive backup of your radio's settings and configurations is imperative. This precautionary step safeguards your preferred frequencies, settings, and configurations in case the update resets the device to its default settings.

4. **Follow Instructions:** Each firmware update is accompanied by specific instructions from Baofeng. These instructions are vital to ensure a smooth update process. Adhering to these guidelines meticulously is crucial to avoid errors during the update and to guarantee the correct installation of the new firmware.

5. **Execute the Update:** Connect your Baofeng radio to a computer using a programming cable that is compatible with your device. Launch the firmware update software and follow the step-by-step instructions provided. Ensure the radio remains connected throughout the update process to prevent interruptions or potential failures.

6. **Verify the Update:** After completing the firmware update, it's crucial to verify that the new firmware has been successfully installed. Check the radio's settings or refer to the update instructions to confirm the updated version. This confirmation assures you that the update process was executed correctly.

7. **Restore Settings:** In case the update reset your radio's settings to default, use the backup created earlier to restore your preferred configurations, frequencies, and settings. This step ensures that your Baofeng radio operates according to your specific preferences post-update.

*Resets*

Resetting Baofeng radios can troubleshoot issues and restore the device to its factory defaults.

**Factory Reset**

Performing a factory reset returns your Baofeng radio to its original factory settings. This action erases all stored frequencies, settings, and configurations. It's advisable to perform a factory reset when troubleshooting persistent issues or when preparing to transfer ownership of the radio.

The exact steps for a factory reset vary among Baofeng models. Typically, it involves pressing specific buttons or following a sequence of steps while turning on the radio. Refer to the user manual or official documentation for the precise steps applicable to your model.

1. **Resetting Specific Settings**

   - **Channels or Frequencies:** Deleting or resetting specific channels or frequencies stored in the radio's memory can resolve issues related to individual settings. Access the radio's menu to remove unwanted frequencies or channels causing problems.
   - **Keypad Lock or Squelch Settings:** Resetting these settings to their default values can troubleshoot problems associated with keypad lock or squelch functionalities, addressing potential functionality issues.

2. **Backup Data Before Reset**

Before executing any reset actions, creating a backup of important settings, frequencies, or configurations is essential. This precaution ensures that you can restore your preferred settings post-reset and avoid the loss of crucial data.

3. **Follow Manufacturer's Instructions**

Adhere strictly to the reset instructions provided by Baofeng in the user manual or official documentation. Different Baofeng models might have slightly different reset procedures, necessitating the use of the correct method for your specific radio.

4. **Verify Reset**

After performing a reset, it's crucial to verify that the changes have taken effect and the Baofeng radio is functioning correctly. Check settings, frequencies, and other configurations to ensure they have been reset or restored accurately.

## Importance of Firmware Updates and Resets

- **Optimizing Performance:** Firmware updates enhance the functionality and performance of Baofeng radios, ensuring they operate efficiently and effectively.
- **Troubleshooting:** Resets help resolve issues that may arise with the radio, such as software glitches, configuration errors, or problematic settings.
- **Staying Updated:** Regular firmware updates keep the radio compatible with new technologies, improve features, and address potential vulnerabilities or bugs.
- **Restoring Defaults:** Factory resets restore the Baofeng radio to its default settings, eliminating persistent issues or errors caused by incorrect configurations.

# Chapter 9: **Safety and Emergency Communications**

## Using Baofeng Radios in Emergencies

During emergencies, Baofeng radios play a crucial role in establishing communication channels when traditional means are unavailable or disrupted. Their portability, versatility, and adaptability make them valuable assets in various emergency scenarios, including natural disasters, search and rescue operations, and critical situations.

**Benefits of Baofeng Radios in Emergencies**

1. **Communication Reliability:** In emergency situations, reliable communication is crucial for coordinating response efforts, ensuring the safety of individuals, and obtaining vital information. Baofeng radios provide a direct line of communication between individuals or groups, allowing for real-time communication without relying on cellular networks or infrastructure that may be compromised during emergencies. They can operate on various frequency bands, including VHF (Very High Frequency) and UHF (Ultra High Frequency), providing flexibility and options for communication.

2. **Long Range Communication:** Baofeng radios offer a relatively long communication range, especially when compared to consumer-grade walkie-talkies. Depending on the model and conditions, Baofeng radios can transmit and receive signals over several miles, extending the reach of communication within a localized area. This extended range is particularly useful in emergency situations where individuals or response teams may be spread out or need to communicate across distances.

3. **Programmable Channels and Frequencies:** Baofeng radios allow users to program and store multiple channels and frequencies, enabling easy access to different communication channels. This feature is valuable during emergencies when different response teams, organizations, or individuals need to communicate on specific channels or frequencies. Users can set up dedicated channels for emergency services, public safety organizations, or community-based communication networks.

4. **Dual Band Operation:** Many Baofeng radio models offer dual band functionality, allowing simultaneous communication on two different frequency bands. For example, a Baofeng radio may operate on both VHF and UHF bands, providing increased flexibility and compatibility with different communication systems. Dual band operation enables users to monitor multiple channels or frequencies simultaneously, enhancing situational awareness during emergencies.

5. **Emergency Alert Features:** Some Baofeng radios include emergency alert features, such as built-in alarms or SOS functions. These features can be activated to signal distress or call for immediate assistance from nearby individuals or response teams. The emergency alert functionality enhances safety and can help expedite rescue or aid efforts during critical situations.

## Best Practices for Using Baofeng Radios in Emergencies

- **Familiarize Yourself with Local Regulations:** Before using Baofeng radios in an emergency, familiarize yourself with local regulations regarding radio usage. Understand the permitted frequency bands, power limits, and any licensing requirements. Adhering to these regulations ensures responsible and legal use of the radios.

- **Obtain the Necessary Licenses:** In some regions, the use of certain frequency bands or high-power transmissions may require obtaining appropriate licenses. If you plan to operate Baofeng radios on licensed frequency bands, ensure that you have the necessary licenses or certifications. Unlicensed or unauthorized use of certain frequency bands can interfere with critical communications and may be subject to legal consequences.

- **Program and Test Your Radios in Advance:** Prior to an emergency, program your Baofeng radios with relevant channels, frequencies, and settings. Test the radios in different conditions and environments to ensure they are functioning properly. Familiarize yourself with the radio's features, controls, and menus to be able to operate them effectively during high-stress situations.

- **Establish Communication Protocols:** Define clear communication protocols and practices for your group or team. Determine which channels or frequencies will be used for specific purposes, such as emergency alerts, coordination, or information dissemination. Establish communication procedures, such as call signs, check-ins, and reporting formats, to ensure efficient and effective communication during emergencies.

- **Maintain Battery Power:** Keep spare batteries or alternative power sources, such as portable chargers or solar chargers, to ensure a continuous power supply for your Baofeng radios. During emergencies, power sources may be limited or unavailable, so it is essential to have backup options to keep your radios operational.

- **Maintain Radio Etiquette:** Practice proper radio etiquette to ensure clear and concise communication. Use clear and standardized language, avoid unnecessary chatter, and wait for pauses before transmitting to avoid interrupting ongoing conversations. Be aware that others may be listening, so maintain professionalism and avoid sharing sensitive or personal information over the radio.

- **Monitor Local Emergency Channels:** Stay informed by monitoring local emergency channels, such as weather alert frequencies or public safety frequencies. These channels may provide important updates, warnings, or instructions from official emergency management agencies or first responders.

- **Coordinate with Local Authorities:** If you are part of an organized emergency response team or community-based network, coordinate your communication efforts with local authorities. Understand their preferred channels or frequencies and collaborate closely to ensure seamless integration and effective communication during emergencies.

- **Practice Emergency Drills:** Conduct regular emergency drills or simulations to practice using Baofeng radios in realistic scenarios. This allows you to familiarize yourself with the radios, test your communication protocols, and identify any areas for improvement. Practice drills can help streamline communication during actual emergencies, reducing response time and enhancing overall effectiveness.

- **Stay Informed and Updated:** Stay informed about the latest developments in emergency communication practices and technologies. Explore additional resources, such as online forums, user groups, or training materials, to expand your knowledge and skills in using Baofeng radios effectively during emergencies. Regularly update your radios' programming and firmware as new features or bug fixes become available from the manufacturer.

- **Respect Privacy and Security:** Keep in mind that radio transmissions are not inherently secure and can be intercepted by others. Avoid transmitting sensitive or personal information over the radio. Exercise caution when discussing operational details or sensitive matters, and use appropriate encryption or coding techniques if necessary.

- **Consider Supplementary Communication Methods:** While Baofeng radios are valuable tools, they should not be relied upon as the sole means of communication during emergencies. Explore other communication methods, such as cellular phones, satellite phones, or alternative radio systems, to establish redundancy and ensure reliable communication in different scenarios.

## Choosing the Right Frequency

Choosing the right frequency is crucial when using Baofeng radios in emergencies. The frequency you select will determine the range, coverage, and compatibility of your communication. In this section, we will explore key factors to consider when choosing the appropriate frequency for your Baofeng radio during emergency situations.

## 1. Understand Frequency Bands

Baofeng radios operate on various frequency bands, including the Very High Frequency (VHF) and Ultra High Frequency (UHF) bands. Each band has its advantages and considerations:

- **VHF Band (136-174 MHz):** VHF frequencies are suitable for communication over relatively long distances in open areas. They provide better coverage and propagation characteristics in rural or outdoor environments. However, VHF signals can be more susceptible to obstacles such as buildings and foliage.
- **UHF Band (400-520 MHz):** UHF frequencies are typically better at penetrating obstacles and are suitable for communication in urban or built-up areas. UHF signals are less affected by buildings and other obstructions, making them advantageous for communication in densely populated areas.

## 2. Local Frequency Regulations

Familiarize yourself with local frequency regulations and licensing requirements in your region. In many countries, Baofeng radios operate within the amateur radio bands, which require an amateur radio license. Ensure you comply with these regulations and understand the permitted power levels, frequency ranges, and other limitations.

## 3. Emergency Service Frequencies

Identify and program emergency service frequencies relevant to your location. These frequencies can include those used by police, fire departments, search and rescue teams, and other emergency response agencies. Having access to these frequencies can allow you to monitor official communications and receive critical updates during emergencies.

## 4. Simplex vs. Repeaters

Baofeng radios support simplex and repeater operations. Simplex communication refers to direct communication between radios on the same frequency. This mode is suitable for short-range communication within a localized area. On the other hand, repeaters are devices that receive signals on one frequency and then retransmit them on another frequency, extending the range of communication. Repeaters are particularly useful for long-range communication or when there are obstacles between radios.

## 5. Interference and Congestion

Consider the potential for interference and congestion on certain frequencies. During emergencies, many individuals and groups may be using radios, leading to crowded frequency bands. In such cases, it is advisable to select less congested frequencies or coordinate with other users to avoid interference and ensure efficient communication.

## 6. Pre-Programmed Emergency Frequencies

Some Baofeng radio models come with pre-programmed emergency frequencies or channels. These frequencies are commonly used for emergency communication and are often included in the radio's memory. Check the user manual or manufacturer's documentation to identify if your radio has pre-programmed emergency frequencies and how to access them.

## 7. Local Amateur Radio Clubs and Networks

Connect with local amateur radio clubs and networks in your area. These organizations often have established emergency communication plans and can provide guidance on recommended frequencies and procedures during emergencies. Collaborating with experienced amateur radio operators can enhance your communication capabilities and overall effectiveness.

## 8. Terrain and Environmental Factors

Consider the terrain and environmental conditions in your area. Mountains, hills, dense forests, and urban structures can significantly impact signal propagation. VHF signals may struggle to penetrate through obstacles, while UHF signals can offer better performance in urban environments. Conduct range tests and assess the local terrain to determine the most suitable frequency for your specific situation.

## 9. Antenna Considerations

The type and quality of your antenna can affect the performance and range of your Baofeng radio. Higher-gain antennas can enhance signal strength and improve communication range. Consider using external antennas or upgrading to aftermarket antennas to optimize your radio's performance in specific frequency bands or environmental conditions.

## 10. Power Output

Baofeng radios typically offer adjustable power output settings. Higher power settings can extend the communication range but consume more battery power. Lower power settings conserve battery life and are suitable for shorter-range communications. Consider the range you need to cover and the available power supply when selecting the appropriate power output for your communication needs.

### 11. Monitoring and Scanning

During emergencies, it is important to monitor different frequencies and channels to stay informed about evolving situations. Baofeng radios often have scanning capabilities that allow you to scan through multiple channels or frequencies to listen for incoming transmissions. This feature enables you to monitor emergency service frequencies, local news channels, or designated emergency communication channels.

### 12. Adaptability and Flexibility

Emergencies can be dynamic, and communication needs may change as the situation evolves. Ensure your Baofeng radio is programmable and adaptable to different frequencies and channels. This flexibility allows you to adjust your communication strategy based on changing circumstances or coordination requirements.

## Accessing NOAA Weather Channels

The NOAA Weather Radio (NWR) stands as a vital nationwide network of radio stations, tirelessly broadcasting continuous and up-to-date weather information directly from the nearest National Weather Service (NWS) office. Operating 24/7, the NWR serves as an indispensable source for timely weather alerts, warnings, forecasts, and various emergency information. Accessing NOAA weather channels on Baofeng radios opens up a gateway to this invaluable resource, significantly enhancing users' ability to stay informed and make well-informed decisions in the face of dynamic weather conditions.

**Steps to Access NOAA Weather Channels on Baofeng Radios**

**Select the Frequency**

- Baofeng radios are designed to operate on specific frequencies. NOAA weather channels, in particular, typically utilize frequencies ranging from 162.400 to 162.550 MHz. To determine the frequency range supported by your specific Baofeng model, consult the user manual for accurate information.

**Enter the Frequency Manually**

- Initiate the process by pressing the "VFO/MR" button on your Baofeng radio, thereby entering the frequency mode.

- Utilize the numeric keypad to manually input the specific NOAA weather channel frequency. For example, for NOAA channel 7, you would enter 162.475 MHz.
- Confirm the entered frequency by pressing the "MENU" button.

## Adjust the Squelch

- The squelch control serves a crucial role in filtering out unwanted noise during periods of no signal. Fine-tune the squelch settings until a clear and audible broadcast is achieved.

## Save the Frequency as a Channel

- Once successfully tuned in to a NOAA weather channel, take the proactive step of saving it as a dedicated channel for convenient future access.
- Navigate to the channel storage settings by pressing the "MENU" button, and then proceed to save the current frequency for easy retrieval.

## Monitoring NOAA Broadcasts

- Post-saving the NOAA weather channel, return to the main screen on your Baofeng radio. From there, seamlessly switch between regular channels and NOAA channels as needed.
- Regularly monitor NOAA broadcasts to stay abreast of real-time updates on weather conditions, forecasts, and crucial emergency alerts.

## Importance of NOAA Weather Channels

1. **Emergency Preparedness:** NOAA weather channels stand as a frontline resource for delivering real-time information on severe weather events, natural disasters, and other emergencies. This immediacy empowers users to take timely and potentially life-saving precautions.

2. **Outdoor Activities:** Outdoor enthusiasts, including hikers, campers, and boaters, find immense value in NOAA weather channels. These channels provide crucial information, allowing individuals to stay informed about changing weather conditions in their specific locations and plan their activities accordingly.

3. **Public Safety:** The accessibility of NOAA weather channels is pivotal for public safety. It ensures that individuals and communities receive immediate alerts and warnings issued by the National Weather Service, fostering a proactive approach to safeguarding lives and property.

4. **Disaster Response:** During catastrophic events such as hurricanes, tornadoes, or floods, NOAA weather channels play a central role in delivering up-to-date information. This

information is instrumental for emergency responders and the general public alike, facilitating well-informed decision-making in critical moments.

## Emergency Codes and Signals

During emergencies, clear and effective communication is paramount. Emergency codes and signals serve as standardized communication tools that convey critical information swiftly and efficiently. Understanding these codes, signals, and their applications is essential for coordinated responses and ensuring safety in crisis scenarios.

### Significance of Emergency Codes and Signals

- **Rapid Communication:** Codes and signals allow for quick transmission of crucial information, enabling responders to act promptly and efficiently during emergencies.
- **Universal Understanding:** Standardized codes and signals ensure a common language for communication, facilitating coordination among various response teams and agencies involved.
- **Minimizing Miscommunication:** Clear and concise codes and signals reduce the chances of misinterpretation or misunderstanding, enabling precise communication in high-stress situations.

### Common Emergency Codes and Their Meanings

1. **Code Blue:** Code Blue is one of the most widely recognized emergency codes used in healthcare settings, particularly in hospitals. It is typically announced over a public address system and signifies a medical emergency, specifically a cardiac arrest. When a Code Blue is activated, healthcare providers rush to the location specified in the announcement to initiate immediate resuscitation efforts.

2. **Code Red:** Code Red is often used to indicate a fire emergency. It alerts staff and patients to the presence of a fire and prompts them to take appropriate action, such as evacuating the area, following designated escape routes, or activating fire suppression systems. In some cases, Code Red may also be used to indicate other types of emergencies, such as a bomb threat.

3. **Code Yellow:** Code Yellow is commonly used in healthcare facilities to indicate a patient who has become agitated, aggressive, or violent. It alerts staff to the need for additional assistance and security measures to ensure the safety of both the patient and the healthcare providers. Code Yellow is often accompanied by specific protocols for handling the situation and de-escalating the patient's behavior.

4.  **Code Black:** Code Black typically refers to a situation involving a bomb threat or the presence of an explosive device. It prompts immediate action to evacuate the area, report the threat to authorities, and implement appropriate security measures. Code Black may also be used in some healthcare settings to indicate a severe weather event, such as a tornado or severe storm.

5.  **Code Orange:** Code Orange is commonly used to signal a mass casualty incident or a hazardous material spill. It signifies a situation where a large number of casualties are expected or where there is a significant risk of exposure to hazardous substances. When Code Orange is activated, emergency response teams are mobilized, and appropriate protocols for patient triage, treatment, and decontamination are implemented.

6.  **Code Silver:** Code Silver is often used to indicate an active shooter situation or the presence of a person with a weapon on the premises. It prompts a rapid response from security personnel and law enforcement agencies to neutralize the threat and ensure the safety of staff, patients, and visitors. Code Silver protocols may include procedures for lockdown, sheltering in place, or evacuating the area.

7.  **Code Gray:** Code Gray is typically used in healthcare settings to indicate a severe weather emergency, such as a tornado, hurricane, or severe storm. It prompts staff and patients to take appropriate shelter and safety measures, such as moving to designated safe areas, securing equipment, and preparing for potential power outages or infrastructure damage.

8.  **Code Amber:** Code Amber is often used to indicate a missing or abducted child. It triggers a coordinated response involving law enforcement agencies, security personnel, and the public to locate and recover the child safely. Code Amber alerts may include detailed descriptions of the missing child, the suspected abductor, and any relevant information that can aid in their recovery.

9.  **Code Green:** Code Green is commonly used to indicate an evacuation or relocation of patients or staff due to a non-fire-related emergency, such as a gas leak, chemical spill, or structural instability. It prompts the activation of evacuation plans, the coordination of transportation, and the relocation of individuals to safe areas or alternative facilities.

10. **Code White:** Code White is often used in healthcare settings to indicate a violent or aggressive person with a weapon. It prompts a rapid response from security personnel and law enforcement agencies to contain the situation, protect staff and patients, and neutralize

the threat. Code White protocols may include lockdown procedures, securing entrances and exits, and providing immediate medical assistance to any injured individuals.

11. **Other Emergency Signals:** In addition to emergency codes, various emergency signals or symbols are used to convey critical information quickly and universally. Some common examples include:

- **Sirens:** Sirens are used by emergency vehicles, such as ambulances, police cars, and fire trucks, to alert others of their presence and the need to give way.
- **Emergency Broadcast System (EBS):** The EBS is a system used to broadcast important emergency information, such as severe weather warnings, public safety alerts, or evacuation orders, through radio, television, or other communication channels.
- **Air Raid Siren:** Air raid sirens are used to signal an imminent attack or threat, particularly during wartime or in areas prone to natural disasters.
- **Strobe Lights:** Strobe lights are often used in emergency situations to attract attention and indicate the presence of a hazard or the need for evacuation.
- **Whistles:** Whistles are commonly used by emergency response personnel, lifeguards, and others to quickly alert and direct people during emergencies, such as in search and rescue operations or during evacuations.

## Communication Protocols

Communication protocols are standardized procedures and practices that ensure efficient and effective communication between individuals or teams. These protocols help streamline information exchange, minimize errors, and facilitate coordination during emergencies. Here are some essential communication protocols to consider when using Baofeng radios:

1. **Radio Check:** Before engaging in critical communication, it is crucial to perform a radio check to ensure that all radios are functioning properly and signal strength is sufficient for reliable communication. A radio check involves a brief exchange to confirm audibility and signal strength between two or more radios. It can be as simple as asking, "Radio check, over," and waiting for a response. If there are any issues with audio quality or weak signals, troubleshooting steps can be taken or alternative communication methods can be used.

2. **Call Signs and Identifiers:** Assigning and using call signs or identifiers for individuals or teams is important for clarity and accountability during radio communication. Call signs can be alphanumeric codes, names, or designations that uniquely identify each user. They should be short, easy to remember, and distinct to avoid confusion. When initiating

communication, use the call sign of the intended recipient to direct the message specifically to them, such as "Team Alpha, this is Team Bravo, over."

3. **Clear and Structured Messages:** To ensure effective communication, messages should be clear, structured, and concise. Use straightforward language and avoid unnecessary jargon or acronyms that may not be universally understood. Structure messages by following a standardized format, such as the Situation-Background-Assessment-Recommendation (SBAR) method. This format includes providing a concise situation overview, relevant background information, an assessment of the current status, and specific recommendations or requests.

4. **Repeat and Confirm:** To minimize misunderstandings and errors, it is important to repeat and confirm critical information during radio communication. After transmitting an important message, ask the recipient to repeat or confirm the information received. For example, "Team Charlie, this is Team Delta, requesting confirmation of our rendezvous point. Over." The recipient should respond by repeating the relevant details or confirming the message accuracy.

5. **Acknowledge and Respond:** Actively acknowledging and responding to received messages is essential for maintaining effective communication. After receiving a message, acknowledge it promptly with a brief response, such as "Copy that" or "Roger." This lets the sender know that their message was received and understood. If a response is required, provide the necessary information or action promptly.

6. **Use Prowords and Procedures:** Prowords (procedure words) are standardized words or phrases used to convey specific meanings or instructions during radio communication. They help streamline communication and ensure common understanding. Some common prowords include "Roger" (affirmative), "Wilco" (will comply), "Negative" (no or not possible), and "Out" (end of transmission). Familiarize yourself with these prowords and use them appropriately to enhance communication clarity.

7. **Situational Updates and Reports:** Regular situational updates and reports are crucial for maintaining situational awareness and facilitating coordinated responses. Designate specific times or intervals for providing updates on the status of the emergency, ongoing activities, resource availability, and any changes in circumstances. Reports should be concise and focus on essential information relevant to the emergency response.

8. **Emergency Procedures:** Establish clear and well-defined emergency procedures that outline specific actions, communication protocols, and roles and responsibilities during various emergency scenarios. These procedures should be communicated to all relevant parties and practiced through drills and exercises. In times of emergency, adhere to the established procedures to ensure a coordinated and efficient response.

9. **Mutual Aid and Interoperability:** In larger emergencies involving multiple response teams or agencies, it is important to establish mutual aid agreements and interoperability protocols. These protocols enable different organizations to communicate and coordinate effectively, even if they are using different radio systems or frequencies. Establish communication channels or liaison officers to facilitate information sharing and ensure seamless collaboration between teams.

10. **Record Keeping and Documentation:** Maintaining accurate records and documentation of radio communications is essential for accountability, analysis, and post-incident evaluation. Record important messages, significant events, decisions, and actions taken during the emergency response. This information can provide valuable insights for future improvements and serve as a reference for legal or investigative purposes if needed.

Remember that effective communication protocols are only as good as the people using them. Regular training, drills, and exercises are crucial to ensure that all individuals involved in emergency response are familiar with the protocols and can execute them confidently and efficiently. Practice scenarios that simulate realistic emergency situations to enhance preparedness and test the effectiveness of communication protocols.

## Tips for Signal Resilience

Maintaining robust communication is crucial in emergencies, and signal resilience plays a pivotal role in ensuring uninterrupted communication. Strategies aimed at enhancing signal resilience with Baofeng radios are imperative to overcome challenges and maintain effective communication amidst adverse conditions.

Signal resilience refers to a system's ability to maintain a consistent and reliable signal transmission even under adverse conditions or external interference. Several factors can impact signal resilience, including terrain, environmental conditions, equipment quality, and interference.

**Factors Influencing Signal Resilience**

1. **Terrain and Topography**

- **Open Areas vs. Urban Environments:** Terrain plays a crucial role in signal transmission. While open areas facilitate better signal propagation, urban environments with buildings and obstructions may hinder transmission.
- **Foliage and Obstacles:** Dense foliage or physical obstacles such as mountains, buildings, or structures can attenuate radio signals, affecting their strength and clarity.

2. **Environmental Conditions**

- **Weather Impact:** Adverse weather conditions like rain, snow, or storms can weaken radio signals, reducing their range and reliability.
- **Atmospheric Conditions:** Variations in atmospheric conditions, such as ionospheric disturbances, can affect the propagation of higher frequency bands, particularly in HF transmissions.

3. **Equipment and Antenna Quality**

- **Antenna Design and Placement:** The quality and positioning of antennas significantly impact signal transmission. Proper antenna design and placement help optimize signal strength and range.
- **Radio Equipment Quality:** The quality and capability of the Baofeng radio itself, including its modulation techniques and power output, affect signal resilience.

**Strategies for Signal Resilience Enhancement**

In this section, we will explore several tips to enhance signal resilience when using Baofeng radios in emergencies:

1. **Antenna Orientation and Elevation:** The orientation and elevation of the radio's antenna can significantly impact signal strength and reliability. For optimal performance, ensure that the antenna is fully extended and positioned vertically. Avoid holding the radio close to your body or obstructing the antenna with your hand, as it can weaken the signal. Additionally, if you are in a fixed location, consider elevating the antenna by placing it on a higher surface or using an external antenna for better line-of-sight communication. External antennas are designed to enhance signal strength and improve reception and transmission capabilities. They can be mounted on vehicles, buildings, or portable antenna masts to overcome signal obstacles and improve overall signal resilience.

2. **Choose the Right Channel or Frequency:** Selecting the appropriate channel or frequency is critical for signal resilience. Baofeng radios typically offer multiple channels and frequencies to choose from. In emergency situations, use channels or frequencies that are designated for emergency communications or have been pre-determined in your communication plan. These channels are typically monitored by relevant authorities and have less interference from non-emergency users, improving signal reliability.

3. **Avoid Signal Interference**: Signal interference can disrupt communication and weaken the signal strength. To minimize interference, keep the radio away from other electronic devices that may emit electromagnetic interference (EMI), such as mobile phones, power lines, or large metal objects. Additionally, avoid transmitting near sources of radio frequency interference (RFI), such as electrical substations or industrial equipment, as they can affect signal quality.

4. **Power Management:** Maintaining sufficient battery power is crucial for signal resilience. Regularly check the battery level of your Baofeng radio and have spare batteries or alternative power sources readily available. In emergencies, conserve battery power by minimizing unnecessary transmissions, reducing the radio's transmit power level if possible, and turning off non-essential features such as backlighting or scanning functions.

5. **Use Headsets or External Microphones:** Using headsets or external microphones can help improve signal clarity and reduce background noise during communication. These accessories help direct the microphone closer to your mouth and minimize interference, resulting in clearer transmissions. Ensure that the headsets or external microphones are compatible with your Baofeng radio model and properly connected.

6. **Conduct Range Testing:** Regularly conduct range testing to determine the effective communication range of your Baofeng radio in different environments and conditions. This testing helps you understand the limitations of your radio and identify areas where signal strength may be weaker. By knowing the range of your radio, you can plan your communication strategy accordingly and make informed decisions during emergencies.

7. **Backup Communication Methods:** While Baofeng radios are valuable communication tools, it is essential to have backup communication methods in case of signal failure or other unforeseen circumstances. Consider carrying alternative communication devices such as mobile phones, satellite phones, or signal flares. These backup methods can provide alternative means of communication when the radio signal is weak or unavailable.

8. **Regular Maintenance and Inspection:** Perform regular maintenance and inspection of your Baofeng radio to ensure optimal performance. Clean the contacts, antenna connector, and battery contacts regularly to remove dirt or corrosion that may affect signal transmission. Inspect the antenna for any damage or wear and replace it if necessary. Regularly update the firmware of your Baofeng radio to benefit from any performance improvements or bug fixes released by the manufacturer.

9. **Stay Updated on Weather Conditions:** Weather conditions can impact signal propagation and signal resilience. Stay updated on weather forecasts, especially during severe weather events, as they can affect the strength and reliability of radio signals. In adverse weather conditions, consider adjusting your communication strategy, such as using repeaters or increasing transmit power if necessary, to maintain effective communication. However, keep in mind that higher transmit power consumes more battery power and may not be necessary if you are already within range of the receiving station. Adjust the transmit power level based on the distance and line-of-sight conditions to optimize signal resilience.

## Search and Rescue Operations

Search and rescue (SAR) operations encompass a wide array of missions aimed at locating and aiding individuals in distress or missing persons across diverse terrains and environments. The success of these operations relies significantly on effective communication, swift decision-making, and coordinated efforts among SAR teams.

Baofeng radios stand out as invaluable communication tools in SAR missions due to their portable design, durability, and versatile frequency range. Their compact nature ensures mobility without compromising communication capabilities, enabling SAR teams to maintain connectivity in remote or challenging terrains where conventional communication methods may falter.

Here are some important considerations and strategies for utilizing Baofeng radios effectively in search and rescue operations:

### 1. Pre-Planning and Preparedness

Before initiating SAR operations, it is essential to have a well-defined plan and be prepared. Consider the following:

- Establish communication protocols: Define clear communication protocols that outline call signs, channels, frequencies, and procedures to ensure efficient and effective communication among SAR teams.

- Assign roles and responsibilities: Designate individuals for specific roles such as incident commander, communication officer, field teams, and support personnel. Clearly communicate roles and responsibilities to ensure smooth coordination.
- Conduct training and drills: Regularly train SAR personnel on radio operation, communication protocols, search techniques, and safety procedures. Conduct realistic drills to test communication systems and coordination.

## 2. Frequency and Channel Selection

Baofeng radios offer multiple frequency bands and channels. Select frequencies and channels that are authorized for SAR operations and are least likely to experience interference. Coordinate with local authorities or SAR organizations to identify dedicated SAR frequencies or channels that are widely used in your area.

## 3. Establish Base Stations

Set up a central base station as a hub for communication and coordination. The base station should have a powerful antenna and be located in a central area with good coverage. Assign a dedicated communication officer to manage the base station and monitor all radio traffic. The base station can relay information, assign tasks to field teams, and provide updates to all involved parties.

## 4. Field Team Communication

Field teams play a vital role in SAR operations. Ensure that each field team is equipped with Baofeng radios for communication. Consider the following:

- Assign unique call signs: Provide each field team with a unique call sign or identifier to facilitate clear and efficient communication. Call signs should be short, easy to remember, and distinct to avoid confusion.
- Establish check-in procedures: Develop check-in procedures for field teams to report their status at regular intervals or when reaching specific milestones. This helps maintain situational awareness and ensures the safety of all team members.
- Use predetermined codes or signals: Establish codes or signals to convey specific messages or situations efficiently. For example, a "Code Red" may indicate an immediate emergency, while a "Code Green" may indicate a successful find or resolution.

## 5. Maintain a Log

Maintain a detailed log of all radio communications during SAR operations. This log should include the time, call signs, message content, and any significant information exchanged. The log serves as a valuable reference for post-operation analysis, documentation, and potential investigations.

### 6. Signal Resilience and Range

Signal resilience is critical during SAR operations, especially in challenging terrain or remote areas. Implement the tips for signal resilience outlined in the previous section to maximize signal strength and coverage. Consider using repeaters or establishing relay points to overcome obstacles and extend communication range if necessary.

### 7. Establish Communication Deadlines

During SAR operations, time is often of the essence. Establish communication deadlines for field teams to report their progress or findings. This ensures that timely updates are provided and allows for reassessment of search strategies if required.

### 8. Utilize Cross-Banding

Cross-banding involves using two Baofeng radios on different frequency bands simultaneously. This technique allows for communication between different radio systems or agencies operating on separate bands. Cross-banding can facilitate communication between SAR teams using different radio equipment or frequencies, enhancing interoperability and coordination.

### 9. Conduct Briefings and Debriefings

Before and after SAR operations, conduct briefings and debriefings to ensure effective communication and learning. Briefings should cover objectives, assignments, communication protocols, and safety considerations. Debriefings provide an opportunity to review the operation, identify lessons learned, and make improvements for future SAR missions.

### 10. Coordinate with External Agencies

In complex SAR operations, coordination with external agencies such as law enforcement, fire departments, or aerial support teams may be necessary. Establish communication channels and liaise with these agencies to exchange information, coordinate efforts, and request additional resources or support when needed.

### 11. Consider GPS and Mapping

Integrate Global Positioning System (GPS) devices and mapping tools into SAR operations. Baofeng radios may have GPS capabilities or can be paired with external GPS devices. These tools can enhance navigation, provide accurate location information, and assist in coordinating search efforts.

## 12. Safety Considerations

Prioritize safety during SAR operations. Communicate safety instructions, hazards, and emergency procedures to all involved personnel. Establish a backup communication plan in case of radio malfunctions or loss of signal. Ensure that personnel are trained in emergency response and have access to alternative communication methods such as mobile phones or satellite phones.

## 13. Regular Communication Updates

Maintain regular communication updates with the incident commander, base station, and other field teams. Provide updates on search progress, findings, obstacles, and resource needs. Timely and accurate information sharing is crucial for effective coordination and decision-making during SAR operations.

## 14. Maintain Professionalism and Clarity

During SAR operations, it is important to maintain professionalism and clarity in radio communications. Speak clearly, use standardized terminology, and avoid jargon or ambiguous language. This ensures that messages are understood accurately and minimizes the potential for miscommunication.

## 15. Post-Operation Evaluation

After completing SAR operations, conduct a thorough post-operation evaluation. Review the effectiveness of communication systems, identify areas for improvement, and incorporate lessons learned into future SAR plans and training programs. Regularly update and refine communication protocols based on feedback and operational experiences.

## 16. Legal and Regulatory Compliance

Ensure compliance with local regulations, licensing requirements, and radio usage laws when operating Baofeng radios during SAR operations. Familiarize yourself with any restrictions or limitations imposed by relevant authorities and obtain necessary permissions or licenses for radio operations, if required.

# Chapter 10: **Staying Safe and Legal**

Baofeng radios are popular handheld devices commonly used for communication purposes. While they offer convenience and versatility, it is essential to understand and adhere to licensing requirements, operate within the law, and be aware of potential penalties and enforcement mechanisms to ensure safe and legal usage. This chapter will provide an overview of these key aspects in relation to Baofeng radios.

## Licensing Requirements

Licensing requirements play a crucial role in ensuring the safe and responsible operation of Baofeng radios. Licensing serves as a regulatory mechanism to ensure that users have the necessary knowledge, skills, and permissions to operate these devices.

### 1. FCC Licensing (United States)

The Federal Communications Commission (FCC) is the regulatory authority responsible for managing and licensing radio communication in the United States. When it comes to Baofeng radios, there are specific licensing requirements that users need to comply with. The following licenses are commonly relevant:

- **Technician Class License:** The Technician Class License is the entry-level license for amateur radio operation in the United States. It grants users operating privileges on most amateur radio frequencies. To obtain this license, individuals are required to pass a written examination that covers basic regulations, operating procedures, and electronics fundamentals. The examination is administered by Volunteer Examiner Coordinators (VECs) authorized by the FCC. Once the examination is passed, applicants can submit the necessary paperwork and fees to the FCC for license issuance.
- **General Class License:** The General Class License is an intermediate-level license that provides expanded operating privileges on additional frequency bands compared to the Technician Class License. It requires passing a more comprehensive examination that covers a wider range of topics, including advanced regulations, operating practices, and electronics theory. Similar to the Technician Class License, applicants must submit the required paperwork and fees to the FCC for license issuance.
- **Amateur Extra Class License:** The Amateur Extra Class License is the highest level of amateur radio license in the United States. It grants the most extensive operating privileges across all amateur radio frequency bands. Obtaining an Amateur Extra Class License requires passing a more challenging examination that covers advanced technical and

regulatory topics. Once the examination is successfully completed, applicants can submit the required paperwork and fees to the FCC for license issuance.

It is important to note that each license class has its own set of operating privileges and restrictions. License holders must adhere to the authorized frequency bands and operating practices specified by their license class.

## 2. Ofcom Licensing (United Kingdom)

In the United Kingdom, the Office of Communications (Ofcom) is responsible for regulating the use of Baofeng radios and other communication devices. The licensing requirements in the UK may vary depending on the purpose and frequency bands used. The following licenses are commonly relevant:

### Business Radio (BR) License

The Business Radio (BR) License is required for commercial and non-amateur use of Baofeng radios in the UK. This license authorizes users to operate on specific frequencies allocated for business and professional purposes. The license is issued for a specific location and may have coverage limitations based on the geographical area. To obtain a BR License, individuals or organizations need to apply to Ofcom and provide necessary details such as the intended use, frequency requirements, and location. Ofcom evaluates the application and, upon approval, issues the license along with any associated conditions or restrictions.

### Amateur Radio License

For individuals interested in amateur radio operation in the UK, various license classes are available. These licenses grant specific frequency privileges and require different levels of examination. The license classes include:

- **Foundation License:** This entry-level license grants limited operating privileges and requires passing a multiple-choice examination covering basic regulations, operating procedures, and technical knowledge.
- **Intermediate License:** The Intermediate License provides expanded operating privileges compared to the Foundation License. It requires passing a more comprehensive examination that covers a broader range of topics, including advanced regulations, operating practices, and electronics theory.
- **Full License:** The Full License offers the highest level of operating privileges for amateur radio operation in the UK. It requires passing a more advanced examination that covers a wide range of technical and regulatory topics.

Applicants for amateur radio licenses in the UK must complete the necessary examinations, provide the required documentation, and submit an application to Ofcom. Upon approval, the license is issued along with any specific conditions or restrictions.

### 3. Licensing Requirements in Other Countries

Licensing requirements for Baofeng radios may vary across different countries. It is crucial for users to research and comply with the specific regulations and licensing requirements in their respective countries. Some countries may have specific licenses for amateur radio operation, while others may require general communication licenses for business or personal use. It is recommended to consult the regulatory authority or telecommunications agency in the respective country to obtain accurate information regarding licensing requirements, application procedures, and associated privileges and restrictions.

### Steps to Obtain a License

### Study and Preparation

- **Resources**: Aspiring amateur radio operators use study materials, online resources, practice exams, and books to prepare for licensing examinations. Understanding radio theory, regulations, and practical aspects of radio operations is crucial.
- **Local Clubs and Mentors**: Engaging with local amateur radio clubs or mentors provides guidance, support, and practical experience. Many clubs offer training sessions and practice exams to help candidates prepare.

### Examination Process

- **Scheduling Exams**: Once ready, candidates schedule examinations with authorized testing centers or organizations. Exams typically consist of multiple-choice questions covering various aspects of radio operation and regulations.
- **Passing the Exam**: Successful completion of the exam demonstrates the candidate's understanding of radio theory, regulations, and operating procedures, qualifying them for the respective license class.

### License Application

- **Application Submission**: After passing the exam, candidates submit license applications to the regulatory body. Applications include personal information, examination results, and any required fees.

- **License Issuance**: Upon approval, the regulatory authority issues the amateur radio license. The license document outlines the operator's privileges, callsign, permitted frequencies, and operating conditions.

## Operating within the Law

Operating within the law is paramount when using Baofeng radios or any radio communication equipment. Adhering to regulations ensures responsible and lawful usage, preventing interference with other communications and maintaining the integrity of radio frequencies.

### Regulatory Compliance and Baofeng Radios

#### 1. Legal Framework and FCC Regulaions

Baofeng radios fall under the purview of the Federal Communications Commission (FCC) regulations in the United States. Understanding FCC regulations is crucial for lawful operation. The FCC allocates specific frequency bands for different services, including amateur radio, commercial use, public safety, and more.

#### 2. Authorized Frequency Bands

- **License Classifications:** Individuals holding amateur radio licenses must operate within designated frequency bands corresponding to their license class. Each class grants specific privileges and access to certain frequencies.
- **Avoiding Unauthorized Bands:** Operating outside allocated frequency bands can lead to violations and interference with other radio services. Compliance with allocated bands helps prevent such issues.

### Protocols and Operating Procedures

#### 3. Adherence to Operational Protocols

- **Standardized Practices:** Users must follow established protocols and procedures while operating Baofeng radios. These protocols include using proper identification callsigns, following transmission etiquette, and adhering to language standards.
- **Maintaining Communication Etiquette:** Proper adherence to communication protocols ensures clear and effective communication, particularly in emergency situations.

#### 4. Responsible Spectrum Usage

- **Respecting Spectrum Integrity:** Operating within designated frequency bands prevents interference with licensed transmissions and ensures fair access to radio frequencies for all users.

- **Avoiding Disruptions:** Unauthorized transmissions or operations in unauthorized bands can disrupt critical communication systems, affecting emergency services and other legitimate radio users.

## Compliance and Legal Implications

### 5. Penalties for Non-Compliance

- **FCC Enforcement:** The FCC has the authority to enforce regulations and penalize non-compliance. Penalties can range from warnings, fines, to license revocation for serious violations.
- **Legal Ramifications:** Non-compliance may extend beyond FCC actions and lead to legal consequences under federal laws, resulting in fines, confiscation of equipment, or legal prosecution.

### 6. Enforcement and Reporting

- **Regulatory Oversight:** The FCC monitors radio transmissions for compliance with regulations. Reports from individuals, agencies, or automated systems help identify violations and enforce regulations.
- **Collaboration with Law Enforcement:** Collaboration between regulatory bodies and law enforcement agencies ensures effective enforcement and compliance with radio regulations.

## Importance of Legal Adherence

### 1. Preserving Communication Systems

- **Preventing Interference:** Operating within legal bounds ensures that radio frequencies remain clear, preventing interference and disruptions in critical communication systems.
- **Maintaining Spectrum Integrity:** Adhering to regulations preserves the integrity of allocated frequency bands, ensuring reliable communication for all users.

### 2. Ethical Responsibility

- **Respecting Others' Rights:** Following regulations demonstrates respect for other users' rights and ensures fair access to radio frequencies without disruption.
- **Emergency Preparedness:** Adhering to legal protocols ensures that emergency frequencies are accessible during critical situations, aiding emergency responders and safeguarding public safety.

### Educational Initiatives and Compliance Measures

1. **Regulatory Resources**

- **FCC Guidance:** The FCC provides comprehensive information, forms, and guidelines regarding amateur radio regulations, compliance, and updates on its website.
- **Educational Materials:** Organizations like the American Radio Relay League (ARRL) offer study materials and resources to educate users about FCC regulations and compliance.

2. **Training and Awareness**

- **Educational Programs:** Amateur radio clubs and online platforms conduct courses and training sessions to educate users on operating within legal boundaries.
- **Promoting Compliance:** Encouraging awareness through educational campaigns and practice guidelines fosters a culture of compliance and responsible radio usage.

## Penalties and Enforcement

Understanding the penalties and enforcement related to operating Baofeng radios within legal boundaries is crucial for ensuring compliance with regulations. Let's delve into the nuances of penalties and enforcement mechanisms associated with radio operation:

1. **Penalties**

Non-compliance with licensing requirements and operating regulations can result in penalties. The severity of penalties may vary depending on the jurisdiction and the nature of the violation. The following are potential penalties that can be imposed:

### Fines

Fines are a common form of penalty for violating licensing requirements and regulations. Regulatory authorities have the power to impose monetary fines on individuals or organizations found to be operating Baofeng radios without the necessary licenses or in violation of operating restrictions. The amount of the fine may vary depending on the jurisdiction and the specific violation. Repeat offenses or more serious violations may result in higher fines.

### Confiscation of Equipment

In some cases, regulatory authorities have the authority to confiscate Baofeng radios or other communication equipment used for illegal purposes. Confiscation serves as a deterrent and helps prevent further non-compliant use of the equipment. Confiscated equipment may be retained by the regulatory authority or may be returned to the owner after the necessary penalties have been imposed.

**Other Repercussions**

In addition to fines and equipment confiscation, there may be other repercussions for non-compliant use of Baofeng radios. These can include:

- **License Revocation:** If a licensed user is found to be in repeated violation of regulations or engages in serious misconduct, their license may be revoked. License revocation can result in the loss of operating privileges and may require reapplication and reexamination to regain licensing status.
- **Legal Action:** In some cases, serious violations or intentional misconduct related to the use of Baofeng radios may result in legal action. Legal action can involve civil lawsuits or criminal charges, depending on the nature and severity of the violation. Legal consequences can include fines, penalties, or even imprisonment.

It is important for Baofeng radio users to understand and comply with licensing requirements and regulations to avoid potential penalties and other legal consequences.

## 2. Enforcement Mechanisms

Regulatory authorities employ various enforcement mechanisms to ensure compliance with licensing requirements and regulations. These mechanisms help monitor and control the use of Baofeng radios and address non-compliant activities. Some common enforcement mechanisms include:

**Monitoring and Detection**

Regulatory authorities often employ monitoring and detection techniques to identify illegal or improper use of Baofeng radios. This may involve dedicated monitoring stations, spectrum analyzers, or other tools to detect unauthorized transmissions or interference with licensed services. Monitoring and detection mechanisms help regulatory authorities identify violations and take appropriate enforcement actions.

**Reporting Violations**

Members of the public who observe illegal or suspicious radio activities can report them to the relevant regulatory authorities. Reporting violations helps authorities identify non-compliant users and take necessary enforcement actions. Individuals can report violations through designated channels provided by the regulatory authority or by contacting local law enforcement agencies.

## Cooperation with Authorities

Baofeng radio users are expected to cooperate with regulatory authorities during investigations and enforcement actions. This includes providing required documentation, responding to inquiries, and complying with any requests for information. Cooperation with authorities is essential in ensuring a fair and thorough enforcement process.

## Collaboration with Law Enforcement

Regulatory authorities often collaborate with local law enforcement agencies to enforce licensing requirements and regulations. This collaboration ensures that enforcement actions are carried out effectively and that legal consequences are appropriately addressed. Law enforcement agencies may assist in investigations, provide legal support, and help with the execution of penalties or legal actions.

# Conclusion

Throughout this comprehensive guide, we have delved into the multifaceted world of Baofeng radios, uncovering their diverse functionalities, applications, and the essential role they play in modern communication systems. Baofeng radios have gained a reputation for their versatility, user-friendly interface, and robust features, making them indispensable tools across various industries, emergency scenarios, and enthusiast communities.

Baofeng radios have evolved beyond their initial purpose as communication devices. They now embody reliability, adaptability, and innovation, fostering seamless communication across diverse sectors and scenarios. With a wide range of models catering to different needs, Baofeng radios have become stalwarts of communication technology in fields such as emergency services requiring swift and coordinated responses and enthusiasts exploring the realms of amateur radio.

This guide has taken you through essential aspects, starting from the unboxing and familiarization process with Baofeng radios. We have delved into understanding the basic components, safety precautions, and fundamental concepts of radio communication. Additionally, we have explored emergency codes and signals, highlighting how Baofeng radios facilitate effective communication during critical situations, ensuring swift responses and coordinated actions.

It is important to recognize that Baofeng radios are more than just tools; they represent a gateway to connectivity. They enable individuals, teams, and communities to stay linked, informed, and empowered. Whether it is in public safety, emergency response, outdoor adventures, amateur radio, or community service, Baofeng radios empower users to communicate effectively, exchange vital information, and coordinate actions.

In conclusion, Baofeng radios are not just a brand or a device; they are a testament to the evolution and innovation within the realm of communication technology. They embody reliability, versatility, and adaptability, serving as integral components in establishing effective communication networks across various domains.

As technology advances and communication needs continue to evolve, Baofeng radios remain at the forefront, ensuring connectivity, reliability, and empowerment in every situation. Their impact extends beyond the devices themselves, shaping communication strategies, emergency response systems, and enthusiast communities worldwide.

Baofeng radios stand as symbols of connectivity, reliability, and empowerment in the realm of communication technology. They bridge distances, fostering collaboration, and empowering individuals and communities to communicate effectively, regardless of the situation. Whether they are utilized for professional use, emergency scenarios, or recreational activities, Baofeng radios offer a reliable and versatile communication solution that keeps people connected and safe. With their user-friendly interface, robust features, and application versatility, Baofeng radios have become essential tools in today's interconnected world.

www.ingramcontent.com/pod-product-compliance
Lightning Source LLC
Chambersburg PA
CBHW080949290526

45795CB00009B/2946